CHILDREN'S GEOGRAPHIES

Children's Geographies presents an overview of a rapidly expanding area of cutting-edge research. Drawing on original research in Europe, North and South America, Africa and Asia, the book analyses children's experiences of playing, living and learning.

The diverse case studies range from an historical analysis of gender relations in nineteenth-century North American playgrounds, through children's experiences of after-school care in contemporary Britain, to street cultures amongst homeless children in Indonesia at the end of the twentieth century. Threaded through this empirical diversity, is a common engagement with current debates about the nature of childhood.

The individual chapters draw on contemporary sociological understandings of children's competence as social actors. In so doing they not only illustrate the importance of such an approach to our understandings of children's geographies, they also contribute to current debates about spatiality in the new social studies of childhood.

The Editors: Sarah L. Holloway is Lecturer in Human Geography at Loughborough University; she is co-author of *Geographies of New Femininities*. **Gill Valentine** is Professor of Geography at the University of Sheffield; her numerous publications include co-authoring *Consuming Geographies*, *Cool Places*, and *Mapping Desire*, all published by Routledge.

CRITICAL GEOGRAPHIES
Edited by Tracey Skelton
Lecturer in International Studies, Nottingham Trent University
and
Gill Valentine
Professor of Geography, The University of Sheffield

This series offers cutting-edge research organised into four themes: concepts, scale, tranformations and work. It is aimed at upper-level undergraduates, research students and academics, and will facilitate inter-disciplinary engagement between geography and other social sciences. It provides a forum for the innovative and vibrant debates which span the broad spectrum of this discipline.

CHILDREN'S GEOGRAPHIES

Playing, living, learning

Edited by
Sarah L. Holloway and Gill Valentine

London and New York

First published 2000
by Routledge
11 New Fetter Lane, London EC4P 4EE

Simultaneously published in the USA and Canada
by Routledge
29 West 35th Street, New York, NY 10001

Routledge is an imprint of the Taylor & Francis Group

Typeset in Perpetua by
Florence Production Ltd, Stoodleigh, Devon.
Printed and bound in Great Britain by
Biddles Ltd, Guildford and King's Lynn

British Library Cataloguing in Publication Data
A catalogue record for this book is available
from the British Library

Library of Congress Cataloguing in Publication Data
Holloway, Sarah, 1970–
Children's geographies / Sarah Holloway & Gill Valentine.
p. cm. – (Critical geographies)
Includes bibliographical references and index.
1. Children. 2. Human geography. I. Valentine, Gill, 1965–
II. Title. III. Series.
HQ767.9 .H65 2000
305.23–dc21   99–056246

ISBN 0–415–20729–0 (hbk)
ISBN 0–415–20730–4 (pbk)

To our Mums and Dads, for giving us our childhoods

CONTENTS

CONTENTS

PLATES

MAPS

TABLES

NOTES ON CONTRIBUTORS

Stuart Aitken is Professor of Human Geography at San Diego State University. His research interests are in Urban and Social Geography with an emphasis on families and communities, children and youth, and film. He is author of *Family Fantasies and Community Space* (Rutgers University Press, 1998) and *Putting Children in their Place* (Association of American Geographers, 1994).

Nicola Ansell is Lecturer in Human Geography at Keele University. Her research interests include geographies of gender, education and empowerment in Southern Africa. Her Ph.D. concerns the impact of secondary schooling on rural girls in Lesotho and Zimbabwe.

John Barker is a Research Officer in the Department of Geography at Brunel University. His most recent work has been with Fiona Smith, looking at the provision of out-of-school childcare. He also has more general interests in the geographies of health and sexualities.

Harriott Beazley has recently completed her Ph.D. in the Department of Human Geography, at the Australian National University. Her thesis was an analysis of the geographies and identities of street children's subcultures in Yogyakarta, Indonesia. She is currently a Visiting Scholar at the Centre for Family Research at the University of Cambridge, where, together with Judith Ennew, she is co-editing a textbook on street children with the working title *Shifting Paradigms: Innovations in research and practice with 'street children'*.

Nick Bingham completed his Ph.D. at the University of Bristol. He then worked as a Researcher at the University of Sheffield before taking up a lectureship in Human Geography at the Open University. His research interests lie in the fields of children's geographies, social studies of technologies, and the environment.

Michael G. Bradford is Professor of Geography at the University of Manchester. He has published widely in the fields of geographical education, service provision and social exclusion. From 1996 to 1998, he jointly directed The Business of Children's Play project with Dr John McKendrick (part of the ESRC's Children 5–16 Research Programme).

Pia Christensen is a Research Fellow and Associate Director of the Centre for the Social Study of Childhood, Hull University. Her main interests and publications are in the anthropological study of children's everyday lives and of children's health, with a particular focus on the individual and collective actions of children. Her most recent research (with A. James and C. Jenks) is a study of the perception, understanding and social organisation of children's time, a project funded by the ESRC under the Children 5–16 Research Programme.

Anna V. Fielder is a Research Associate at the Centre for Urban Policy Studies, University of Manchester (UK). She has inter-disciplinary research interests that span the fields of geography, politics and health care. From 1996 to 1998, Anna was employed as Research Associate on The Business of Children's Play project with Professor Michael Bradford and Dr John McKendrick. She is part of the ESRC's Children 5–16 Research Programme.

Shaun Fielding is a Research Manager with EcoTec Research and Consultancy where he conducts and develops programme evaluations for the public and commercial sectors. Previously he was a Research Fellow in the School of Education at Birmingham University where he worked on an ESRC project examining the different strategies for learning adopted by boys and girls in primary schools (with colleagues in Birmingham and London). He has worked on numerous education research projects in the academic and public sectors, including the restructuring of work in primary schools, lifelong learning in European cities and informal learning and widening participation.

Elizabeth Gagen is a doctoral student in the Department of Geography at Cambridge University. Her Ph.D. research is on the US playground movement, 1880–1920. She has taught and studied as a postgraduate at Syracuse University and the University of Kentucky. Her broad research interests are in identity and space, historical method, and children's geographies.

Sarah L. Holloway is Lecturer in Human Geography at Loughborough University. Her research interests lie in the fields of social, feminist and children's geographies. She has published work on mothering and childcare cultures, as well as children's use of new information and communications technologies. She is co-author (with C. Dwyer, N. Laurie and F.M. Smith) of *Geographies of New Femininities* (Longman, 1999).

Allison James is Reader in Applied Anthropology at the University of Hull and currently Director of the Centre for the Social Study of Childhood. Her main research interests are in childhood, ageing and the life course. Her most recent publications are *Childhood Identities* (Edinburgh University Press, 1993); *Growing Up and Growing Old* (with J. Hockey) (Sage, 1993) and *Theorising Childhood* (with C. Jenks and A. Prout) (Polity Press, 1998). She is currently researching (with P. Christensen and C. Jenks) children's perception and understandings of time on a project funded under the ESRC Children 5–16 Research Programme.

Chris Jenks is Professor of Sociology and Pro-Warden (Research) at Goldsmiths' College, University of London. His most recent major publications are *Childhood* (Routledge, 1996); *Theorising Childhood* (with A. James and A. Prout) (Polity Press, 1998) and *Core Sociological Dichotomies* (Sage, 1998). He is currently researching (with P. Christensen and A. James) children's perception and understandings of time on a project funded under the ESRC Children 5–16 Research Programme.

Owain Jones has an M.A. in Environmental Philosophy and Geography from the University of the West of England; M.Sc., Society and Space, and a Ph.D. from the University of Bristol. He is currently a Research Fellow at the University of Bristol, he is also Visiting Lecturer in Environmental Philosophy and Politics at the University of the West of England.

Lily Kong is Associate Professor at the Department of Geography, National University of Singapore. Her research focuses on religion, music, constructions of 'nation' and constructions of 'nature', particularly in the context of the multi-racial, multi-religious postcolonial city-state of Singapore.

Melanie Limb is Lecturer in Human Geography and Assistant Director of the Centre for Children and Youth at University College Northampton. Her research interests are in the area of children's and young people's geographies, looking in particular at children's use of and values for place and space, and youth participation in local political processes.

Hugh Matthews is Professor in Geography and Director of the Centre for Children and Youth at University College Northampton. His research interests are in the geographies of children – in particular how they make use of place and understand space – and the geography of disability. He is author of *Making Sense of Place: Children's Understanding of Large-Scale Environments* (Harvester Wheatsheaf, 1992) and co-author of *Urban Development and Change* (with Philip Allen, 1998).

John H. McKendrick is Lecturer in the School of Social Sciences at Glasgow Caledonian University. His research interests revolve around the geography of children and family life. Current research includes the commercialisation and privatisation of children's leisure space in the UK (The Business of Children's Play project), and re-engaging non-participating young people in learning activities. He is part of the ESRC's Children 5–16 Research Programme.

Samantha Punch is currently working as Research Fellow on a project Children and Welfare: Negotiating Pathways (part of the ESRC Children 5–16 Research Programme). This study focuses on Scottish children's coping strategies, support networks and help-seeking behaviour in relation to problems in their daily lives. She has also conducted ethnographic research on children's lives in rural Bolivia, exploring how they negotiate their autonomy at home, at school, at work and at play.

Elsbeth Robson is Lecturer in Development Studies at Keele University. Her research interests include gender issues and young people's geographies in Africa. She has conducted fieldwork in Kenya, Nigeria and Zimbabwe. Currently she is researching the experiences of Zimbabwean young carers.

Tracey Skelton is Lecturer in Geography at Nottingham Trent University. Her research interests revolve around the cultural and feminist geographies of young people; gender, race, culture and development issues; and longitudinal research based on the Caribbean island of Montserrat. She is co-editor (with G. Valentine) of *Cool Places* (1998) and (with T. Allen) of *Culture and Global Change* (1999), both published by Routledge.

Fiona Smith is Lecturer in Human Geography at Brunel University. Her research interests include the provision of out-of-school childcare, the changing life of the child and children's use of space. She participated in the ESRC's Children 5–16 Research Programme.

Mark Taylor is a Researcher in the Department of Environmental Sciences at University College Northampton. His research interests lie in the field of children's geographies, particularly children's use of urban space and their involvement in the planning processes which shape the built environment.

Gill Valentine is Professor of Geography at the University of Sheffield, where she teaches social geography and philosophy and methodology. She is co-author (with D. Bell) of *Consuming Geographies* (Routledge) and co-editor (with T. Skelton) of *Cool Places: geographies of youth cultures* (Routledge) and co-editor (with D. Bell) of *Mapping Desire* (Routledge).

ACKNOWLEDGEMENTS

We would like to thank all those who participated in the RGS/IBG session on 'Geographies of childrearing . . . children's and young people's geographies' (Kingston, UK January 1998), from which the idea for this book stemmed. In particular, we want to thank Claire Dwyer for nudging us into action. Thanks are also due to organisers – Stuart Aitken, Doreen Mattingly and Tom Herman – and participants at the Geographies of Young People workshop at San Diego State University for renewing our enthusiasm part way through the editing process. Our thinking about children's geographies has also benefited from participation in the ESRC's Children 5–16 programme, and in particular we would like to acknowledge the support of Allison James, Chris Philo, Alan Prout and the Built Environment Group. We have also benefited from the friendship and academic support of Ruth Butler, Claire Dwyer, Nina Laurie, Tracey Skelton, Fiona Smith and Fiona Smith (!) as well as numerous colleagues in our respective departments during the writing of this book. Finally, thanks to Deborah and Sarah for sending in food parcels when it seemed as though the book might never be finished.

Every attempt has been made to obtain permissions to reproduce copyright material. If any proper acknowledgement has not been made, we would invite copyright holders to inform us of the oversight.

1

CHILDREN'S GEOGRAPHIES AND THE NEW SOCIAL STUDIES OF CHILDHOOD

Sarah L. Holloway and Gill Valentine

Introduction

Children's Geographies presents an overview of a rapidly expanding area of research. Drawing on original research in Europe, North and South America, Africa and Asia, this volume provides an analysis of children's experiences of playing, living and learning. The case studies are diverse, ranging from an historical analysis of gender relations in nineteenth-century North American playgrounds, through children's experiences of after-school care in contemporary Britain, to street cultures amongst homeless children in Indonesia at the end of the twentieth century. Threaded through this empirical diversity is a common engagement with current debates about the nature of childhood. The individual chapters draw on contemporary sociological understandings of children's competence as social actors. In so doing they not only illustrate the importance of such an approach to our understandings of children's geographies, they also contribute to debates about spatiality in the new social studies of childhood. In this chapter we introduce the academic contexts from which the individual studies stem and provide an outline of the rest of the book. The next section discusses the social construction of childhood and the changing approach to the study of children within the social sciences. The subsequent section illustrates the importance of geographical work on children, highlighting the difference that place makes, the importance of the different sites of everyday life and the spatial imagery in ideologies of childhood. We end by providing an outline of the rest of the book, which is divided into three sections – playing, living and learning.

Children and childhood

The invention of childhood

Debates about identity and difference have been a dominant focus of interest in the social sciences during the 1980s and 1990s. A common trend in these rather disparate sets of writings (for example, Hall 1992; Rutherford 1990; Weeks 1985; Young 1990) has been to dispute essentialist assumptions about identity and instead uncover the ways in which our fractured identities are socially constructed. Children and the concept of childhood have a rather interesting relation to these debates. Like many social identities, child appears at first sight to be a biologically defined category, marked in this case by chronological age. Children, it is commonly assumed, are those subjects who have yet to reach biological and social maturity – quite simply they are younger than adults, and have yet to develop the full range of competencies adults possess. On the one hand, this less-than-adult status means that childhood is a time when children are to be developed, stretched and educated into their future adult roles, most clearly through the institution of schooling, but also through the family and wider social and civic life. In the most general sense, then, childhood is understood to be a time of socialisation where children learn what it is to be fully human adult beings. On the other hand, as less-than-adults, children in the West are assumed to have the right to a childhood of innocence and freedom from the responsibilities of the adult world. Thus responsible adults have a duty to protect children from dangerous knowledges and people, and in normal circumstances children are not expected to contribute economically to their households or the care of others.

This marking of children as adults' 'other' is challenged by academics who illustrate that the current understanding of children in the North as being less developed, less able and less competent than adults (Waksler 1991) is historically specific. Most well known is the work of Ariès (1962) who traced the emergence of childhood through an analysis of cultural artefacts (see also Cunningham 1991; Hendrick 1990; Stainton-Rogers and Stainton-Rogers 1992; Steedman 1990). Though drawing mainly on French culture, his analysis is conventionally supposed to be generalisable to the rest of the Western world (Jenks 1996). Ariès (1962) argues that the ancient world's understandings of childhood had been forgotten in medieval civilisation, and that in the Middle Ages young people were regarded as miniature adults, rather than conceptually different from adults. 'Children' were thus not present in medieval icons, and only began to be represented in later centuries as ideas about childhood began to develop. In the sixteenth century children began to emerge as playthings for adults from privileged backgrounds. This marked the beginning of the process through which children came to be understood as separate and distinct types of beings, and with the Enlightenment these more modern conceptions of childhood began to dominate. Jenks (1996: 65) argues that it was in this period that children 'escaped into difference' as the category of the child came to be viewed as inherently different from that of the adult.

Jenks (1996) identifies two ways of thinking and talking about this distinctive category of the child that emerge from the historical and cross-cultural literature, which he names as Dionysian and Apollonian views of childhood. Dionysian understandings of childhood were the first to emerge and in this view, children are conceived of as little devils, as inherently naughty, unruly, unsocialised beings. The strong links between such ideas and Christian doctrine are evident in Schnucker's (1990) analysis of late sixteenth- and early seventeenth-century Puritan childrearing advice. The nineteenth-century Evangelical movement in Britain promoted similar attitudes to childrearing, as evidenced in this extract of a letter from Susanna Wesley to her son John:

> the parent who studies to subdue (self-will) in his children, works together with God in the saving of a soul: the parent who indulges it does the devil's work; makes religion impracticable, salvation unattainable; and does all that in him lies to damn his child, soul and body, for ever! . . . This, therefore, I cannot but earnestly repeat, – Break their wills betimes; begin this great work before they can run alone, before they can speak plain, or perhaps speak at all. Whatever pains it cost, conquer their stubbornness; break the will, if you would not damn the child. I conjure you not to neglect, not to delay this! Therefore (1) Let a child, from a year old, be taught to fear the rod and to cry softly. In order to do this, (2) Let him have nothing he cries for; absolutely nothing, great or small; else you undo your own work. (3) At all events, from that age, make him do as he is bid, if you whip him ten times running to effect it. Let none persuade you it is cruelty to do this; it is cruelty not to do it. Break his will now, and his soul will live, and he will probably bless you to all eternity.
>
> (Wesley 1872, quoted in Newson and Newson, 1974: 56)

This concentration on defeating the 'devil within' needs to be understood in the context of very high infant mortality rates (see Walvin 1982 for examples). Given the widespread belief in Heaven and Hell, preparing a child for death was as important as preparing them for life: a disciplined child might fail physically but would be sure of eternal salvation (Newson and Newson 1974; Richardson 1993).

Apollonian views of childhood emerged later than Dionysian ones, and were formalised in the mid-eighteenth-century works of Rousseau, who celebrated children's natural virtues and talents which adults could develop by gentle coaxing:

If the philosophy of the Enlightenment brought to eighteenth century Europe a new confidence in the possibility of human happiness, special credit must go to Rousseau for calling attention to the needs of children. For the first time in history, he made a large group of people believe that childhood was worth the attention of intelligent adults, encouraging an interest in the process of growing up rather than just the product. Education of children was part of the interest in progress which was so predominant in the intellectual trends of the time

(Robertson 1976: 407, quoted in Jenks 1996: 65)

In this view, children are represented as little 'angels', who are born good and innocent of adult ways. These ideas underlie much of the growth since the late nineteenth century in concern for the education and welfare of children, which is evidenced in the provision or regulation of much childcare, education, and interventionist welfare services. That Apollonian 'angels' emerged after Dionysian 'little devils' should not, however, be taken to mean that the former has now replaced the latter: both apparently contradictory understandings of the child continue to be mobilised in contemporary Western societies (as we discuss later), and neither should be considered unproblematic (Jenks 1996).

In both Apollonian and Dionysian constructions of childhood, children's ('good' or 'bad') behaviour is assumed to be a natural part of what it is to be a child, natural tendencies which must be shaped or curbed by adults. Both conceptions are thus essentialist, seeing 'childish' behaviour as stemming from biological impulses. In illustrating the historical specificity of contemporary understandings of childhood in the North, the authors discussed earlier show that the child, far from being a biological category, is a socially constructed identity. Not only is the category of the child a recent invention, but the qualities supposed to be natural in children have changed over time (and, as we discuss later, space). The category child – like woman, 'white', or disabled – is thus shown to be a social construction which, to follow Connell's (1987: 78) arguments about gender, is 'radically unnatural', being inscribed on the body through a lengthy historical process. Only in 'modern' times has physical immaturity been socially dealt with through the historical process we call childhood (cf. Connell 1987: 79). Unlike most other social identities, childhood is something which all adult beings have experienced rather than a difference which forever separates people. Perhaps this is why children, as adults' 'other', are not feared and loathed (cf. Rose 1993 on masculinity and femininity) but openly valued, with childhood being celebrated for those of an appropriate age. Interestingly, though this conception of childhood is recent, its precise demarcation remains problematic. Official definitions of where one ends and the other begins vary even within individual countries, and these are contested by different groups of children and adults (see Hendrick 1990; James 1986; Sommerville 1982; Stainton-Rogers and Stainton-Rogers 1992; Steedman 1995). It is to the studies of the experience of childhood that we now turn.

The new social studies of childhood

One 'academic' consequence of the construction of the child as less than adult, and childhood as a phase of socialisation, is that research on children has been less valued than that on other topics, and children have in the past been far from visible even in research which concerns their everyday experiences. Ambert (1986), for example, identified a near absence of children in North American sociological research, and argued that this reflected the continuing influence of founding theorists whose preoccupations were shaped by the patriarchal values of the societies in which they lived (and hence paid little attention to children), and the system of rewards within the discipline which favours research on the 'big issues' such as class, bureaucracies, or the political system. Even in those branches of the discipline where children might be expected to feature, for example the sociology of the family or education, they were in the past strangely missing (Brannen and O'Brien 1995; James, Jenks and Prout 1998). Here, children tended to be seen as human becomings rather than human beings, who through the process of socialisation were to be shaped into fully human adult beings. As children in this view were regarded as incompetent and incomplete, as 'adults in the making rather than children in the state of being' (Brannen and O'Brien 1995: 70), it was the forces of socialisation – the family, the school – which received attention rather than children themselves.

Recognition that childhood is a socially constructed phenomenon, however, has been accompanied by a challenge to 'traditional' approaches to the study of children with their emphasis on the socialisation of children through various stages of development (Oakley 1994). Researchers in what are now called the new social studies of childhood provide a two-pronged challenge to the relative absence of children from the sociological research agenda. First, their insistence that childhood is a social construction which varies with time and place and as it articulates with other social differences (Prout and James 1990), forms the basis of a research agenda which places an analysis of the social construction of different childhoods at centre stage. If childhood is a social rather than a biological phenomenon – which varies between social groups, societies and historical periods – its construction, contestation and consequences are worthy of academic attention. Second, the new social studies of childhood claim an epistemological break from previous sociological work, in that they study children as social actors, as beings in their own right rather than as pre-adult becomings (Brannen and O'Brien 1996; James et al. 1998; Qvortrup et al. 1994;

Waksler 1991). Rather than accepting contemporary constructions of children in the North as less able and competent than adults, these researchers insist that children are active beings whose agency is important in the creation of their own life-worlds:

> Children are and must be seen as active in the construction and deter-
> mination of their own social lives, the lives of those around them and
> of societies in which they live. Children are not just the passive subjects
> of social structures and processes
>
> (Prout and James 1990: 8)

Children's competence as social actors is a key theme in the new social studies of childhood. The aim, however, is not to celebrate children's creativity and resourcefulness to the detriment of an analysis of wider social structures. Recognition of children's agency does not necessarily lead to a rejection of an appreciation of the ways in which their lives are shaped by forces beyond the control of individual children. As James and Prout argue:

> A more satisfactory theoretical perspective would be one that could
> account for childhood as a structural feature of society in the moment
> of its impinging upon children's experiences in daily life and the
> reshaping of the institution of childhood by children through their day
> to day activities. In essence, it would address both structure and agency
> in the same movement.
>
> (James and Prout 1995: 81)

Nor, as we suggested earlier, does a recognition of children's agency result in a universalisation of the category 'child'. Though children are defined in rela-tion to adults, other differences also fracture (and are fractured by) these adult–child relations. Children's identities are classed, racialised, gendered and so on, just as gender, class and racialised identities are cross-cut by adult–child relations. Moreover, these adult–child relations are constituted in different ways in different times and places.

The changing name given to such approaches to childhood – from the soci-ology of childhood (Jenks 1982), to the sociological study of childhood (James and Prout 1990), to the new social studies of childhood (James et al. 1998) – is important, reflecting an explicit recognition of the growing cross-fertilisation of ideas between researchers in a variety of social science disciplines, including geography (ibid.). This theoretical framework informs the individual chapters of this book which seek not only to explore the different constructions of childhood in different times and places, but also the important role children themselves play in experiencing, constructing and contesting these definitions. We recognise that to prioritise one axis of identity as the basis for organising a book is not an unproblematic act. Different axes of identity are not simply additive – a subject is not simply a child, a boy and a member of the racialised minority, for example –they are mutually constituting (cf. Brand 1990; Ware 1992, cited in Laurie *et al.* 1999). Here we choose temporarily to prioritise the category 'child' and adult–child relations over other identities and social relations in order to highlight a group who have, for a long time, suffered relative neglect in academic work. In solidifying for a moment the category 'child' we seek not to essentialise or universalise it: throughout the book the contributors aim to uncover the social construction of childhood, and in doing so they draw out differences between subjects defined as children in different times and places. These differences are not simply in terms of other axes of identity such as gender and race, for the category 'child' is itself fractured by the ways bodily development is understood in different times and places. For example, while Stuart Aitken's chapter discusses pre-linguistic children, Tracey Skelton focuses on teenage girls on the verge of adulthood.

Children and geography

Reaching critical mass

In this section we outline the important contribution geographical approaches have and can make to the new social studies of childhood. Children have not been a traditional focus of concern in geography, as in many other social science disciplines. Nevertheless, the efforts of a few key individuals mean that we have a small but significant literature about children's environments which dates back to the 1970s and includes studies of children's spatial cognition and mapping abilities as well as their access to, use of and attachment to space. As Aitken outlines:

> The growing interest in this area may be traced in part from William Bunge's (1973; Bunge and Bordessa, 1975) geographical 'expeditions' in Detroit and Toronto which focused upon the spatial oppression of children. The central thesis of this work posits children as the ultimate victims of the political, social and economic forces which contrive the geography of our built environment . . . At about the same time, James Blaut and David Stea founded the 'Place Perception Project' at Clark University (Blaut and Stea 1971, 1974; Blaut, McCleary and Blaut 1970). Although their work was not as polemical as Bunge's, these two researchers established an agenda for research in children's geographic

learning, generated some new theory and conducted some provocative empirical research on the early mapping abilities of children.

(Aitken 1994: 3–4)

During the late 1970s and 1980s other researchers started to build upon this grounding, making a number of significant contributions (Hart 1979; Matthews 1992; Spencer, Blades and Morsley 1989). However, for the most part this work has been ignored within an adultist discipline, such that James was still able to ask in 1990, 'Is there a place for children in geography?' (James 1990; Sibley 1991; Winchester 1991). The last decade of the twentieth century has seen renewed interest in incorporating children's voices and experiences within the geographical project (Aitken 1994; Matthews and Limb 1999; Philo 1992; Sibley 1995a) and we would argue that the research in children's geographies is now beginning to reach a critical mass. This is evidenced in recent sessions at the annual conferences of the Royal Geographical Society/Institute of British Geographers and the Association of American Geographers, as well as in research projects and seminars funded by government bodies in several countries. In particular, there is a growing body of work on children's geographies in European and Scandinavian countries, though because of language barriers this work is not widely acknowledged by British and North American writers (see, for example, a collection of work from an international congress at the University of Amsterdam edited by Camstra 1997).

The two-fold split evident in work on children's geographies from the early 1970s is still evident today. On the one hand, some researchers continue to draw on and inform psychological interest in children's spatial cognition and mapping abilities (e.g. Blades *et al.*, 1998; Blaut 1991, 1997; Matthews 1987, 1995a; Sowden *et al.* 1996; Stea, Elquea and Blaut 1997). This tradition has critiqued Piagetian models of development through experiments which uncover children's early mapping abilities. On the other hand, a largely different group of researchers draw on sociological interest in children as social actors in furthering Bunge's commitment to give children, as a minority group, a voice in an adultist world. The insistence in this approach that children are compe-tent social actors has led to the development of child-centred methodologies. These allow children to construct accounts of their lives in their own terms – for example by inviting children to help design questionnaire surveys and iden-tify key themes for semi-structured interviews – while also recognising the ethical complexities of working with children – for example the need to obtain informed consent, and researchers' responsibilities towards children (see Alderson 1995; Matthews, Limb and Taylor 1998b; Valentine 1999). This present book draws on and contributes to debates in the second of these two traditions. Geographers' contributions to the psychological tradition have been the subject of numerous volumes (e.g. Matthews 1992; Spencer *et al.* 1989); the aim of this collection is to give a similar platform to work which stems from sociological roots.

Our intention in this section is to provide an overview and prospectus of work in this second field. Rather than produce an extended chronological narrative, or summarise the work according to nuances in the philosophical orientations of different authors, we organise our discussion around different ways of thinking about spatiality (see Holloway and Valentine forthcoming for an extended discussion of some of the issues raised in this section). In so doing, we find the three-fold typology of feminist work put forward by Laurie *et al.* (1999) – which emphasises the importance of place, everyday spaces and spatial discourses – particularly helpful, and borrow this to head our subsections and inform our discussion. Under these headings we discuss geographical work (by geographers and those interested in geographical ideas) which focuses on the ways in which children negotiate the childhoods constructed in various times and places.

The importance of place

Perhaps the most obvious, but also one of the most important contributions that geography can make to the new social studies of childhood is to illustrate the importance of place. In a very general sense, geographical studies can add texture and detail to the currently rather broadbrush analysis of the social construction of childhood. As we demonstrated in the previous section, numerous researchers have challenged essentialist understandings of 'the child' by arguing that childhood is constructed in different ways in different times and places. Though a range of studies have illustrated this point in terms of time, much less consideration has been given to space, with most work continuing to focus on constructions of childhood in the North. A number of geographical studies in the South, however, have started to show that the 'normal' assumptions about childhood with which we began our discussion in the previous section, and which are a common focus of discussion in much empirical work, are far from 'normal' assumptions in many contexts. This point is perhaps best illustrated by example.

The notion that children are dependent on adults and are not, in normal circumstances, expected to contribute to household income or the care of others, has promoted empirical work in the North which shows that this powerful construction is always reflected in practice. Some children are shown to be important care providers for their ill or disabled parents, for example, and this finding has stimulated policy concern about the impacts this has on children's childhoods and appropriate policy interventions (Aldridge and Becker 1993, 1995; Becker, Aldridge and Dearden 1998; Stables and Smith forthcoming). This particular construction of childhood as a time of dependency, which is sometimes lost when children have to take on more adult roles, is not always relevant.

Robson (1996), for example, makes this point in relation to children's role in household survival. Focusing on the youthful societies of West Africa, which have nearly half their populations comprising children under 14, she argues:

> Demographically such populations are said to have high dependency rations, but the notion of children as dependants fails to acknowledge the work children do in contributing to household survival and well-being. Children contribute to household production; for example, in poor rural households, children perform many labour-intensive agricultural tasks which would otherwise be done by adults, probably women.
>
> (Robson 1996: 403)

Children in these households are not considered exceptional in their own countries, as are young carers in the North, rather their activities are an everyday feature of social reproduction. Though there are important questions to be asked about the global distribution of resources (Aitken 1994), which means that some children must work to ensure household survival whilst others can over-consume, we need to balance our concerns for the rights of children with a recognition that 'universal' rights are often based on ethnocentric definitions of childhood. Such a balancing act is a trick not easily performed (see Roberts 1998 for a discussion of these ideas in relation to child labour in export industries of the South).

The emphasis in geographical work on place is not only useful in reminding us that conceptions of childhood are spatially as well as temporally specific. Work which draws on what Massey (1993, 1994) has termed a progressive sense of place also usefully illustrates the connections between global and local processes. James *et al.* (1998) provide a useful classification of (mainly non-geographical) work within the new social studies of childhood. They identify an irreconcilable split between research which is global in its focus – for example that which examines the importance of global processes in shaping children's positions in different societies across the world – and that which has more local concerns – for example studies which show how children are important in creating their own cultures and life-worlds. Work in children's geographies, however, shows that if an alternative and more thoroughly spatial understanding of global/local is employed, the dichotomy is transcended and productive cross-linkages can be formed.

Katz's (1993, 1994) work in New York and the Sudan is perhaps the most innovative in making this point. Her aim is to provide texture to the notion of global change by demonstrating some of its local consequences. By focusing on New York and a village in Sudan, she demonstrates that the local manifestations of global restructuring have had serious, negative consequences for children. In particular, she points to the way global processes cause systematic disruption of social reproduction and highlights its consequences for children in both locations, most notably that they fail to receive the knowledge and skills necessary for their adult futures. For example, cuts in public spending mean that many young New Yorkers are let down by a failing public school system, and that children in Howa, a Sudanese village, are learning farming skills which global shifts in the political ecology of agriculture are rapidly making redundant. In so doing, she does not focus solely on global processes, but also demonstrates how children actively experience and respond to these 'global' changes. For example, in Howa, the environmental degradation brought about by the incorporation of the village into an agricultural development project means that children had to travel further to collect water and wood and to graze animals. In undertaking these tasks, children are not passive victims of global processes but exercise control over their own life-worlds, for example choosing from a young age the terrain they would work, their paths to and from it, and the time at which they would return to the village each day. This study illustrates that 'local' cultures – how children organise their day – are bound up with 'global' processes; and that 'global' processes do not exist in the ether, but are worked out in 'local' places. Global and local studies are not irreconcilably split, rather they intimately bind together.

Everyday spaces

A second, and very much related way in which we may think about spatiality and childhood, is to focus on those everyday spaces in and through which children's identities and lives are made and remade. This perception of spatiality is perhaps the best developed in geography and the new social studies of childhood more generally. James *et al.* (1998) are the group of sociologists and social anthropologists within the new social studies of childhood who have given this particular focus on spatiality most consideration (see also Chapter 9 in this book). Focusing on three sites – the home, the school and the city – they explore 'how each is dedicated to the control and regulation of the child's body and mind through regimes of discipline, learning, development, maturation and skill' (James *et al.* 1998: 38). Here we review the more extensive geographical literature focusing on the different sites of everyday life, pinpointing in particular work on the street, the playground, the school and the home.

Children's geographers' principal concern with the street, and 'public' space more generally, has been to examine children's access to, use of and attachment to space. Several studies have particularly focused on children's play environments. These have charted the uneven geography of specialist facilities, activities and programmes for children (e.g. Hill and Michelson 1981) while also exploring children's play preferences and their ability to carve out opportunities for play irrespective of the level of formal play provision (Moore and Young 1978; Ward 1978, 1990). Most notably, early work by Hart (1979) in an unidentified US town (which he gave the pseudonym Invale) focused on the ways that young children use, experience and value space differently from adults. He particularly noted the ways that children appropriate public space for themselves and give names to their favourite places which reflect the way they use them – such as 'sliding hill'. Subsequently, a number of other studies have observed the amount of time children spend playing in the dirt or snow and constructing their own secret places and dens within 'public' space (Wood 1985, Ward 1990), or turning aspects of the everyday landscape from ornamental ponds to walls into skateboard runs (Aitken and Ginsberg 1988). Indeed, Kevin Lynch's (1997) study of children in a suburban neighbourhood in Melbourne found that when children were denied this opportunity to create their own play space (adults prevented the children he studied from establishing a bike track) they became bored and dissatisfied. As a consequence, researchers such as Ward (1978, 1990) and Sibley (1991) have argued that there is a mismatch between formally designated and provided playgrounds and the more flexible landscapes, such as waste ground and open space, which children actually prefer to play in.

More recent work has addressed growing public concerns in the North about children's presence in 'public' spaces, mostly notably the twin fears that some (Apollonian) children are vulnerable to dangers in 'public' places, and the unruly behaviour of other (Dionysian) children can risk adult control of 'public' space (Valentine 1996a, 1996b). Valentine's study (1997a, 1997b, 1997c) considers how parents and children conceive of and negotiate these risks in socio-economically mixed urban and rural areas, emphasising both the importance of local parenting cultures and children's agency in the construction and contestation of family rules about use of the street. Matthews, Limb and Percy-Smith's (1998) study of a slightly older group of children pays similar attention to children's active use of the street. They argue that the urban street environments of young teenagers are not just appendages to the adult world, but are special places, created by themselves and invested with their own values. Matthews' and Limb's concern for young teenagers' use of the street is also evident in their more recent work which highlights the importance of the street in the lives of boys and girls from working-class backgrounds. One particularly interesting strand of their work is their analysis of children's exclusion from the planning processes which influence the design of the built environment, and their attempts to involve children in formal structures that can counter this (Matthews 1995b; Matthews and Limb 1999; Matthews et al. 1998; Matthews, Limb and Taylor 1998, forthcoming). That this process is unlikely to be a smooth one, is evident from Lees' (1999) work in the US which highlights adults' deliberate attempts to construct land and soundscapes unattractive to children in order to discourage their use of 'public' space.

Work on children's use of 'public' space in the South provides a useful counterpoint to these debates, reminding us again of the difference that place makes. Katz's work (1993, 1994), discussed earlier, signals an important point: though the children in her Sudanese case study were constrained in the chores they had to undertake as part of daily social reproduction, they were spatially much freer than Western children, though girls' mobility was reduced as they reach puberty. This contrasted directly with her New York case study where children's access to 'public' space was more constrained and grew only with age. This point is reinforced by Punch's (1998) work in Bolivia. The children in her study used their spatial freedom in order to gain control over their own time, which, like those in Katz's study, was constrained by the household and agricultural chores in which they were involved. For example, children's work in minding grazing animals took them some distance from their homes, and they could use this distance from their parents to secure time off, for example pretending to look for a lost animal and returning home later than expected. As Punch makes clear in this book, these experiences – rather than those of highly constrained access to 'public' space – are shared by the majority of the world's children.

In the North the development of playgrounds has been, throughout the course of the twentieth century, a common response to fears about children's safety in 'public' space, and concerns for their proper, rather than uncontrolled, development. Gagen's (1998) work on the development of the playground movement in Cambridge, Massachusetts in the early 1900s illustrates the importance of concerns among middle-class Americans about the proper development of growing numbers of immigrant children who were not being 'assimilated' into the national way of life. These concerns resulted in the growth of playgrounds with regimes designed to produce appropriately gendered American citizens. Interest in children's playspaces continues through to the present day, and includes questions about the design of playgrounds which are both safe and enticing for children (see Special Issue of *Built Environment* 1999). Though unsupervised public playgrounds continue to be of great importance in many Northern countries, a newer trend towards the commercialisation of playspace is also in evidence. Focusing on Britain, McKendrick, Fielder and Bradford's (1999) work illustrates the diverse forms these newer playgrounds take, including, for example, commercial indoor soft-play centres and family pubs with children's indoor and outdoor playspaces. McKendrick *et al.* argue that these commercial playspaces raise new questions about children's place in 'public' space. Specifically, the growth of this sector increases the range of places available for children's play within urban environments; however, not all families can afford to use these services, and children's direct access to play space is not enhanced as the decision to use these facilities is more often made by parents than children.

Schools, like playgrounds, are also institutional spaces through which adults attempt to control children, and through which differences between children are reinscribed. Aitken makes this point forcefully:

> A major purpose of school control is to socialize children with regard to their roles in life and their places in society. It serves the larger stratified society by inculcating compliant citizens and productive workers who will be prepared to assume roles considered appropriate to the pretension of their race, class and gender identities.
>
> (Aitken 1994: 90)

Ploszajska's work on Victorian reformatory schools provides a particularly good illustration of Aitken's general argument. She focuses on two reformatories – Redhill farm school for boys and Redlodge for girls – and argues that the 'class dimension of the reformatory system as a whole is unambiguous; reformatory girls and boys were, indeed, trained for work in the service of, and subordinate to, the middle-classes' (Ploszajska 1994: 426). Yet her class analysis is also cross-cut by gender, as the institutional location, design and assumption about employment post-reformatory were also shaped by gendered concerns. For example, a rural school was considered most suitable for boys, a large proportion of whom emigrated to be farm labourers in the colonies after being discharged, whilst a suburban setting was considered more appropriate for girls, who more often entered domestic service on leaving the school. Contemporary work in the USA and the UK also illustrates the importance of schools as sites through which gender and sexual identities are made and remade. Hyams (forthcoming) examines discourses of femininity amongst Latina girls in Los Angeles, showing how ideas about appropriate femininities both structure and are contested through the girls' everyday practices. Similarly Holloway, Valentine and Bingham (forthcoming) show that the reproduction of gender differences in information technology lessons is mediated through the heterosexual economy of the classroom. Work about experiences in the South is less readily available, but Ansell's (1999) recent discussion of the reproduction of gender identities in Southern African schooling systems shows that the importance of schools as sites of control, and as spaces through which differences between children are reproduced, is widespread.

The final 'everyday space' we want to consider is the home. Most geographical work on the home as a 'locus of power relations' (Sibley 1995b: 92) has been conducted by feminist researchers and focuses on questions about gender relations within households headed by heterosexual couples (see for example

England 1996). As a consequence of this important line of enquiry, other members of these families, mainly children but also elders, are constructed in terms of the time/care demands they place upon the household rather than as social actors in their own right. In contrast, Aitken (1994), and Sibley (1995a, 1995b) identify the home as an important site for the negotiation of adult–child power relations. Wood and Beck (1990, 1994) observe that the home is a space which is constituted through familial rules and regulations which demarcate appropriate ways for children to behave. Aitken and Herman (1997) draw on psychoanalytic work to discuss this in relation to very young children, but the remainder of this emerging body of work tends to focus on children in middle or late childhood. Sibley (1995a) draws on mass observation data and adults' childhood reminiscences to examine struggles over space within heterosexual family households. Such intra-household relations are also considered in Holloway and Valentine's (1999) work on children's domestic use of new information and communication technologies (ICT), alongside an analysis of the embedding of the household in local and global networks. In contrast, other work considers households experiencing difficult circumstances; for example, Sibley and Lowe (1992) look at households where a parent abuses alcohol, and Stables and Smith (forthcoming) look at children involved with the care of disabled parents.

Significant attention has also been paid to questions of homelessness, and in particular to the street cultures children actively construct for themselves. Winchester and Costello (1995) focus on the complex cultures of Australian street children which are both resistant to dominant cultures and sometimes highly conservative, for example containing strains of both patriarchal and more egalitarian gender relations. As Winchester and Costello (1995) note, the social networks children form when living in the street, which are often essential to their survival in this space, also provide a self-maintaining force which can repro-duce a culture of homelessness (cf. Ruddick 1996 on homeless youth).

Spatial discourses

The final way in which we want to think about spatiality is distinctly different from, though again related to, the previous two approaches. Specifically, we want to elucidate the links between childhood as a discursive construction and a variety of spatial discourses, including those focusing on the home, the city street, the rural idyll and national identity. Turning first to the home, we can see that childhood, for many in the North at least, has been increasingly domes-ticated over the course of the past two centuries. This process is not simply a material one, in the sense that children are spending increasing amounts of time in the home, but is also ideological, in that there is a sense in which this is where children should spend their time. The importance of the home for very young children is evident in ideas about childrearing in Britain, which in the post-war period have tended to suggest that home-care is best for pre-school-aged children (Richardson 1993). The importance of these ideas is evident in the conscious decision some working mothers make to select childcare arrangements – such as care by other family members, childminders and less often nannies – which reproduce home-style environments for their children. Indeed Gregson and Lowe (1995: 231), in their study of nanny-employing households, go so far as to argue that 'Home space remains in many senses unchallenged as the space for young children in contemporary Britain. Small wonder that locating childcare in the parental home continues to matter so much to the middle classes'.

The social construction of childhood in contemporary Britain is highly spatialised. Like any ideological construction, however, the idea that home is the best place for young children is open to challenge, and Holloway's (Holloway 1998; Laurie *et al.* 1999) work on mothers both in and out of paid employment not only shows the enduring power of home in British childrearing ideologies, but also breakages and reinterpretations of the meaning of home. For example, the growing emphasis on the importance of educating pre-school-aged children alongside the availability of affordable collective childcare is leading some middle-class mothers to challenge the assumption that home is best and instead to place greater value on the stimulation collective-care environments can provide. In this way changing ideas about childhood, in this case that children need a pre-school education, can change the meaning of the home and collective childcare spaces, such that children's place is no longer seen to be always in the home.

The way in which our understandings of childhood shape the meaning, and therefore the use of, particular spaces is also clear when we focus on school-aged children. We discussed earlier how contradictory ideas about children as either angels or devils has produced different concerns about children's ventures out of the home into the street in the North (Valentine 1996a, 1996b). Though the understandings of children as either angels or devils are in some ways contradictory, both 'stories' reproduce the same spatial ideology that children's place is in the home, and in straying outside this they either place themselves at risk in adult controlled space, or their unruly behaviour threatens adult hegemony of public space. These stories are important because responses to them – curfews in extreme circumstances, but more often attempts by parents to encourage their children into more home-based or formally organised events outside the home (Valentine 1997a, 1997b) – reinforce adults' control of 'public' space (see Valentine and Holloway forthcoming for a discussion about parental fears about children's safety in cyberspace, which both draw on and rework these ideas in new ways).

Jones's work on English country childhoods provides a useful contrast to these arguments about children's access to the street. He points out that 'both country and childhood are massively and intricately present within our culture(s). Where they intersect in ideas of "country childhood" they often become a vision of considerable potency' (Jones 1997: 158). The two intersections he highlights are stories about country childhoods (for example recollected rural childhoods such as Laurie Lee's *Cider with Rosie*, and contemporary discourses about rural childhood reproduced through lifestyle magazines) and stories for childhood (including a wide range of books written for children which are set in a rural context, for example Enid Blyton's *Famous Five* books, Arthur Ransome's *Swallows and Amazons* novels, A. A. Milne's *Winnie-the-Pooh* stories). In these stories of and about childhood, children's presence in the country is naturalised: children are portrayed playing outdoors, with companions, beyond the surveillance of adults, blessed through their proximity to and interaction with nature. These are 'angelic' children whose innocence is reproduced through their closeness with and to nature. In the context of the rural idyll, Apollonian conceptions of childhood merge with idealised understandings of the rural, to produce a new subtheme in rural discourse, the rural childhood idyll. As Jones (1997) highlights, this story, or discourse, like those about the home and the street, is potentially powerful, shaping both the decisions of individual parents to raise their children in the country, and a wider policy neglect of the potential difficulties children face in the country (see also Philo 1992; Ward 1990; Valentine 1997c).

Other spatial discourses, and here we particularly want to highlight nationalism, also contain ideas about the meaning of childhood, and shape individual children's lived experiences. Despite increasing interest in nationalist discourses in geography, and the ways they are gendered, racialised and so on (Marston 1990; Radcliffe 1996; Sharpe 1996), little consideration has yet been given by geographers to the place of children in these discourses, or the consequences of these discourses for individual children (but see Gagen 1998 and Chapter 12 in this book for an exception). Work elsewhere in the new social studies of childhood (see *Childhood* 1997) suggests that geographers are currently missing much of interest. Stephens (1997), for example, highlights the centrality of images of children and childhood in the American Cold War consensus:

> Notions of biologically mandated, developmentally normal childhood – and the social relations and practices necessary to ensure them – were foundational elements of a Cold War consensus, because children were seen as the basic and unquestionable 'atoms' of society. Their 'natural' needs for security, stability, clear and firm gender differentiations, and protections from both internal and external dangers called for and legitimated the construction of a complex national defence apparatus, organized according to a 'logic of containment' that spanned the boundaries between private and public, psychological and political spheres.
>
> (Stephens 1997: 112)

Stephens also highlights the material implication of Cold War policies for different groups of children, for example those included unwittingly in radiation experiments, those living near nuclear test sites and, at a more general level, a generation of children across the nation who were taught 'duck and cover' in the event of a nuclear attack. Her point that studies of nationalism need to consider both the imagined place of children in the nation, and the consequences of nationalist discourses for individual children, is an important one. The discontinuity between the two in America, led some mothers to campaign against the nuclear arms race: concerns for 'real' children were used to justify political opposition to a system based on 'imagined' ideas of childhood (Stephens 1997).

Crosslinkages

Our aim in this section has been to provide an overview of recent work within children's geographies. Though not comprehensive, this analysis illustrates three interrelated ways in which geography matters in the social construction of childhood, and to the everyday experiences of children. First, geographical work can be used to counter the danger of ethnocentrism by showing that place matters. These places that matter, the work reviewed here shows, are simultaneously global and local, places where children both experience and rework global processes in creating their own worlds of meaning. Second, this review has highlighted numerous studies that illustrate the ways in which children's identities and lives are made and (re)made through the sites of everyday life. Specific attention has been drawn to the street, the playground, the school and the home, though this is not intended to be an exhaustive list. Finally, this review makes clear that our ideas about childhood and our ideas about different spaces/places inform one another.

In structuring this section around three different focuses on spatiality we are not trying to suggest there are three distinct and different ways in which spatiality matters in the social construction of childhood and children's lived experiences. Rather, we have tried to illustrate the ways in which these three focuses on spatiality are closely intertwined. To reiterate, the spaces of everyday life we have highlighted are produced through their webs of connections within wider global social processes (which in turn are reshaped through their constant rearticulation), just as spatial discourses are important as they inform socio-spatial practices in the spaces of everyday life (which in turn reinforce our spatialised ideas about childhood). As Massey argues:

> the social relations which constitute space are not organised into scales
> so much as into *constellations of temporary coherences* . . . set within a

social space which is the product of relations and interconnections from the very local to the intercontinental.

(Massey 1998: 124–5)

Children's geographies

The remainder of this volume is divided into three sections: playing, living and learning. The contributions collected under the heading *playing* take the debates outlined earlier about children's access to 'public' space as their background and move the debate forward from here. Owain Jones provides texture to his literary account of rural childhoods, in an empirically grounded assessment of children's country childhoods. He outlines the ways in which children construct smooth, or at least differently striated, space out of adult space in the village of 'Allswell' in England. Sam Punch looks at rural childhoods in a very different context, that of rural Bolivia, to show how majority world children's spatial freedom allows them to develop some autonomous use of time for play. Hugh Matthews, Melanie Limb and Mark Taylor draw on their recent empirical study of economically disadvantaged estates to point out that girls' use of the street as an urban playspace is greater in some housing areas than discourses about child dangers, and dangerous (male) children, might lead us to expect. Tracey Skelton picks up this theme in her analysis of working-class girls' use of 'public' space in Wales. The girls in her study escape the confines of home by using the street as a leisure space, or attending a community project which blurs the boundaries of 'public' and 'private' space. In the final contribution in this section, John McKendrick, Mike Bradford and Anna Fielder note the expansion of children's play opportunities in the city through the emergence of commercialised play-spaces and analyse their importance through a focus on children's birthday parties.

The contributions in the section on *living* chart the changing experience of children at the end of the twentieth century. Stuart Aitken employs a partial critique of object relations theory in his analysis of increasing institutionalisation of the early years of childhood. His concern with the quality of parent–child relationships is picked up by Pia Christiansen, Allison James and Chris Jenks, who explore the importance of 'quality time' for primary-school-aged children. They identify the importance of spatiality, both within the home and in movement in and out of the home, to children's experiences and control of 'family' and 'own' time. This concern with intra-household relations is also evident in Gill Valentine, Sarah Holloway and Nick Bingham's discussion of children's domestic use of ICT. They draw on empirical work with children in families to debunk the myths about children's use of the Internet, namely that ICT is addictive, isolates children, and puts them in danger on-line. Elsbeth Robson and Nicola Ansell turn our attention away from learning and leisure in the home through their analysis of caring stories written by a group of children in Zimbabwe, which highlight the important caring work undertaken by children. They not only illustrate who children care for, how, and why, but also explore the importance of Christian doctrine in shaping children's understandings of these experiences. In the final chapter in this section Harriott Beazley examines the cultures of street children in Java, exploring the importance of their multiple experiences of 'home' in shaping children's lives on the street.

The contributions in the final section of the book address the variety of children's *learning* environments. Liz Gagen traces the development of the playground movement in Cambridge, Massachusetts and its attempts to produce appropriately gendered future American citizens. In the process she explores how gender differences were inscripted onto children's bodies through the organisation of playground regimes. Shaun Fielding builds on this and other historical analyses, in his examination of the moral geographies in British primary schools. He highlights the important moralities evident in classroom geographies and ends with an impassioned call for greater collaboration between geographers and education research. Fiona Smith and John Barker shift our attention away from the school *per se*, to out-of-school care. Their concern is to highlight children's experiences of out-of-school club environments, and to emphasise both their capacity to rework club spaces to their own ends as well as barriers to children's control of childcare spaces. In the final chapter of the book, Lily Kong shifts our attention again, this time to nature as an outdoor classroom. She argues that Singaporean children are losing access to this important site of learning in a highly urbanised society, and that policy measures are required to redress this.

Our aim in editing this collection is to contribute to and broaden debates within children's geographies. As we suggested earlier, work in the psychological tradition has already been showcased in a number of books, and our intention here is to foreground work which contributes to the social studies of childhood. We do not imagine that this volume will play a definitive role, but hope that it will become part of the process by which children's geographies as an issue, though not necessary as a sub-disciplinary theme, gain a firmer footing within the mainstream geographical research agenda. Other elements in this process are the numerous special editions in geographical and wider social science journals which are scheduled to come out over the next couple of years, the many conferences which focus on and integrate children's geographies within their sessions, and publications from a wide range of research projects which are currently underway. The outcome of these processes is as yet far from clear, and the question of whether 'children's geographies' should be singled out as a sub-disciplinary interest, or integrated into broader thematic interests remains a moot one. We would argue that collections such as this, which provide a focus within a relatively (re)new(ed) field, are important at one point in time, but would hope, as children's issues are integrated into the geographical agenda, that volumes such as this might be consigned to a particular stage of the discipline's contested history.

References

Aitken, S.C. (1994) *Putting Children in their Place*, Washington, DC: Association of American Geographers.

Aitken, S.C. and Ginsberg, S. (1988) 'Children's characterization of place', *Yearbook of the Association of Pacific Coast Geographers* 50: 67–84.

Aitken, S.C. and Herman, T. (1997) 'Gender, power and crib geography: transitional spaces and potential places', *Gender, Place and Culture* 4, 1: 63–88.

Alderson, P. (1995) *Listening to Children: Children, Ethics and Social Research*, Ilford: Barnardos.

Aldridge, J. and Becker, S. (1993) 'Punishing children for caring: the hidden cost of young carers', *Children and Society* 7, 4: 376–87.

—— (1995) 'The rights and wrongs of children who care', in B. Franklin (ed.) *The Handbook of Children's Rights: Comparative Policy and Practice*, London: Routledge.

Ambert, A. (1986) 'Sociology of sociology: the place of children in North American sociology', in P. Alder and P. Alder (eds) *Sociological Studies of Child Development Vol. 1*, Greenwich, Conn.: JAI Press.

Ansell, N. (1999) 'Southern African secondary schools: places of empowerment for rural girls?', unpublished Ph.D. thesis, Keele University, UK.

Ariès P. (1962) *Centuries of Childhood*, New York: Vintage Press.

Becker, S., Aldridge, J. and Dearden, C. (1998) *Young Carers and their Families,* Oxford: Blackwell.

Blades, M., Blaut, J.M., Darvizeh, Z., Elguea, S., Sowen, S. Soni, D., Spencer, C., Stea, D. Surajpaul, R. and Uttal, D. (1998) 'A cross-cultural study of young children's mapping abilities', *Transactions of the Institute of British Geographers* 23, 2: 269–77.

Blaut, J.M. (1991) 'Natural mapping', *Transactions of the Institute of British Geographers* 16, 1: 55–74.

—— (1997) 'The mapping abilities of young children', *Annals of the Association of American Geographers* 87, 1: 152–8.

Blaut, J.M. and Stea, D. (1971) 'Studies of geographic learning', *Annals of the Association of American Geographers* 61: 387–93.

—— (1974) 'Mapping at the age of three', *Journal of Geography* 73: 5–9.

Blaut, J.M., McCleary, G. and Blaut, A. (1970) 'Environmental mapping in young children', *Environment and Behaviour* 2, 3: 335–49.

Brand, D. (1990) 'Bread out of stone', in L. Scheier, S. Sheard and E. Wachtd (eds) *Language in Her Eye*, Toronto: Coach House Press.

Brannen, J. and O'Brien, M. (1995) 'Childhood and the sociological gaze: paradigms and paradoxes', *Sociology* 29: 729–37.

—— (eds) (1996) *Children in Families: Research and Policy*, London: Falmer Press.

Built Environment (1999) Special issue on 'Playgrounds and the built environment', ed. J.H. McKendrick, 25, 1.

Bunge. W.W. (1973) 'The geography', *Professional Geographer* 25, 4: 331–7.

Bunge, W.W. and Bordessa, R. (1975) *The Canadian Alternative: Survival, Expeditions and Urban Change*, Geographical Monographs No. 2, Toronto: York University.

Camstra, R. (1997) (ed.) *Growing Up in a Changing Urban Landscape*, Assen, The Netherlands: Van Gorcum & Comp.

Childhood (1997) Special Issue on 'Children and nationalism', ed. S. Stephens, 4, 1.

Connell, R.W. (1987) *Gender and Power: Society, the Person and Sexual Politics*, Cambridge: Polity Press.

Cunningham, H. (1991) *The Children of the Poor: Representations of Childhood Since the Seventeenth Century*, Oxford: Blackwell.

England, K. (ed.) (1996) *Who Will Mind The Baby? Geographies of Childcare and Working Mothers*, London: Routledge.

Gagen, E.A. (1998) '"An example to us all": children's bodies and identity construction in early twentieth century playgrounds', Geographies of Young People and Young People's Geographies Workshop, San Diego State University, November 1998.

Gregson, N. and Lowe, M. 1995. '"Home"-making: on the spatiality of daily social reproduction in contemporary middle-class Britain', *Transactions of the Institute of British Geographers* 20: 224–35.

Hall, S. (1992) 'New ethnicities', in J. Donald and A. Rattansi (eds) *'Race', Culture and Difference*, London: Sage.

Hart, R. (1979) *Children's Experience of Place*, New York: Irvington.

Hendrick, H. (1990) 'Constructions and reconstructions of British childhood: an interpretative survey, 1800 to present', in A. James and A. Prout (eds) *Constructing and Reconstructing Childhood: Contemporary Issues in the Sociological Study of Childhood*, London: Falmer Press.

Hill, F. and Michelson, W. (1981) 'Towards a geography of urban children and youth', in D. Herbert and R. Johnston (eds) *Geography and the Urban Environment*, New York: John Wiley.

Holloway, S.L. (1998) 'Local childcare cultures: moral geographies of mothering and the social organization of pre-school children', *Gender, Place and Culture*, 5, 1: 29–53.

Holloway, S.L. and Valentine, G. (forthcoming) 'Spatiality and the new social studies of childhood', *Sociology: The Journal of the British Sociological Association*.

—— (1999) 'Multi-media(ted) homes: children, parents and the domestic use of ICT', paper available from the authors, Department of Geography, Loughborough University, Loughborough, Leicestershire LE11 3TU, UK.

Holloway, S.L., Valentine, G. and Bingham, N. (forthcoming) 'Institutionalised technologies: masculinities, femininities and the heterosexual economy of the IT classroom', *Environment and Planning A*.

Hyams, M. (forthcoming) '"Pay attention in class . . . [and] don't get pregnant": a discourse of academic success among adolescent Latinas', *Environment and Planning A*.

James, A. (1986) 'Learning to belong: the boundaries of adolescence', in A.P. Cohen (ed.) *Symbolising Boundaries: Identity and Diversity in British Cultures*, Manchester: Manchester University Press.

James, A. and Prout, A. (1995) 'Hierarchy, boundary and agency: toward a theoretical perspective on childhood', in A. Ambert (ed.) *Sociological Studies of Children* 7: 77–101, Greenwich, Conn.: JAI Press.

—— (eds) (1990) *Constructing and Reconstructing Childhood: Contemporary Issues in the Sociological Study of Childhood*, Basingstoke: Falmer Press.

James, A., Jenks, C. and Prout, A. (1998) *Theorizing Childhood*, Cambridge: Polity Press.

James, S. (1990) 'Is there a place for children in geography?' *Area* 22, 3: 278–83.

Jenks, C. (ed.) (1982) *The Sociology of Childhood – Essential Readings*, London: Batsford.

—— (1996) *Childhood*, London: Routledge.

Jones, O. (1997) 'Little figures, big shadows: country childhood stories', in P. Cloke and J. Little (eds) *Contested Countryside Cultures: Otherness, Marginalisation and Rurality*, London: Routledge.

Katz, C. (1993) 'Growing girls/closing circles: limits on the spaces of knowing in rural Sudan and US cities', in C. Katz and J. Monk (eds) *Full Circles: Geographies of Women Over the Life Course*, London: Routledge.

—— (1994) 'Textures of global changes: eroding ecologies of childhood in New York and Sudan'. *Childhood: A Global Journal of Childhood Research* 2: 103–10.

Laurie, N., Dwyer, C., Holloway, S.L. and Smith, F.M. (1999) *Geographies of New Femininities*, Harlow, Essex: Longman.

Lees, L. (1999) 'The emancipatory city: any space for youth?' paper presented at the Association of American Geographers Annual Meeting, Hawaii, March.

Lynch, K. (1997) *Growing up in Cities*, Cambridge, Mass.: MIT Press.

Marston, S.A. (1990) 'Who are "the people"?: gender, citizenship and the making of the American nation', *Environment and Planning D: Society and Space* 8: 449–58.

Massey, D. (1993) 'Power geometry and a progressive sense of place', in J. Bird, B. Curtis, T. Putnam, G. Robertson and L. Tickner (eds) *Mapping the Futures: Local Cultures, Global Change*, London: Routledge.

—— (1994) *Space, Place and Gender*, Cambridge: Polity Press.

—— (1998) 'The spatial construction of youth cultures', in T. Skelton and G. Valentine (eds) *Cool Places: Geographies of Youth Cultures*, London: Routledge.

Matthews, H. (1987) 'Gender, home range and environmental cognition', *Transactions of the Institute of British Geographers* 12: 43–56.

—— (1992) *Making Sense of Place: Children's Understandings of Large-Scale Environments*, Hemel Hempstead: Harvester Wheatsheaf.

—— (1995a) 'Culture, environmental experience and environmental awareness: making sense of young Kenyan children's views of place', *Geographical Journal* 161, 3: 285–95.

—— (1995b) 'Living on the edge: children as "outsiders"', *Tijdschrift voor Economische en Sociale Geografie* 86, 5: 456–66.

Matthews, H. and Limb, M. (1999) 'Defining an agenda for the geography of children: review and prospect', *Progress in Human Geography* 23, 1: 61–90.

Matthews, H., Limb, M. and Percy-Smith, B. (1998) 'Changing worlds: the micro-geographies of young teenagers', *Tijdschrift voor Economische en Sociale Geografie* 89, 2: 193–202.

Matthews, H., Limb, M. and Taylor, M. (1998a) 'The right to say: the development of youth councils/forums in the UK', *Area* 30, 66–78.

—— (1998b) 'The geography of children: some ethical and methodological considerations for project and dissertation work', *Journal of Geography in Higher Education* 22, 3: 311–24.

—— (forthcoming) 'Young people's participation and representation in society', *Geoforum*.

Matthews, H., Limb, M., Harrison, L. and Taylor, M. (1998) 'Local places and the political engagement of young people: youth councils as participatory structures', *Youth and Policy* 62, Winter: 16–30.

Mattingly, D. and the Geographies of Young People Workshop (1999) 'Growing up: geographic research on children and teenagers comes of age', paper available from the authors, Departments of Geography and Women's Studies, San Diego State University, USA.

McKendrick, J.H., Fielder, A.V. and Bradford, M.G. (1999) 'Privatisation of collective play spaces in the UK', *Built Environment* 25, 1: 44–57.

Moore, R.C. (1986) *Childhood's Domain: Place and Play in Child Development*, London: Croom Helm.

Moore, R.C. and Young, D. (1978) 'Childhood outdoors: towards a social ecology of the landscape', in I. Altman and J. Wolhwill, *Children and the Environment: Advances in Theory and Research*, New York: Plenum Press.

Newson, J. and Newson, E. (1974) 'Cultural aspects of childrearing in the English-speaking world', in M.P.M. Richards (ed.) *The Integration of a Child into a Social World*, London: Cambridge University Press.

Oakley, A. (1994) 'Women and children first and last: parallels and differences between women's and children's studies', in B. Mayall (ed.) *Children's Childhoods: Observed and Experienced*, London: Falmer Press.

Philo, C. (1992) 'Neglected rural geographies: a review', *Journal of Rural Studies* 8, 2: 193–207.

Ploszajska, T. (1994) 'Moral landscapes and manipulated spaces: gender, class and space in Victorian reformatory schools', *Journal of Historical Geography* 20, 4: 413–29.

Prout, A. and James, A. (1990) 'A new paradigm for the sociology of childhood? Provenance, promise and problems', in A. James and A. Prout (eds) *Constructing and Reconstructing Childhood*, London: Falmer Press.

Punch, S. (1998) 'Children's strategies for controlling their use of time and space in rural Bolivia', paper presented at the Royal Geographical Society with the Institute of British Geographers Annual Conference, 5–8 January, University of Kingston, UK.

Qvortrup, J., Bardy, M., Sgritta, G. and Wintersberger, H. (eds) (1994) *Childhood Matters: Social Theory, Practice and Politics*, Aldershot: Avebury.

Radcliffe, S.A. (1996) 'Gendered nations: nostalgia, development and territory in Ecuador', *Gender, Place and Culture* 3, 1: 5–21.

Richardson, D. (1993) *Women, Motherhood and Childrearing*, London: Macmillan.

Roberts, S. (1998) 'Child labour: geographies of connection and context', paper presented at the Geographies of Young People and Young People's Geographies Workshop, San Diego State University, November.

Robertson, P. (1976) 'Home as nest: middle class childhood in nineteenth century Europe', in L. DeMause (ed.) *The History of Childhood*, London: Souvenir.

Robson, E. (1996) 'Working girls and boys: children's contribution to household survival in West Africa', *Geography* 81: 403–7.

Rose, G. (1993) *Feminism & Geography: The Limits of Geographical Knowledge* Cambridge: Polity Press.

Ruddick, S.M. (1996) *Young and Homeless in Hollywood: Mapping Social Identities*, London: Routledge.

Rutherford, J. (ed.) (1990) *Identity: Community, Culture, Difference*, London: Lawrence and Wishart.

Schnucker, R.V. (1990) 'Puritan attitudes towards childhood discipline, 1560–1634', in V. Fildes (ed.) *Women as Mothers in Pre-industrial England*, London: Routledge.

Sharp, J.P. (1996) 'Gendering nationhood: a feminist engagement with national identity', in N. Duncan (ed.) *BodySpace: Destabilising Geographies of Gender and Sexuality*, London and New York: Routledge.

Sibley, D. (1991) 'Children's geographies: some problems of representation', *Area* 23, 3: 269–70.

—— (1995a) 'Families and domestic routines: constructing the boundaries of childhood', in S. Pile and N. Thrift (eds) *Mapping the Subject: Geographies of Cultural Transformation*, London: Routledge.

—— (1995b) *Geographies of Exclusion*, London: Routledge.

Sibley, D. and Lowe, G. (1992) 'Domestic space, modes of control and problem behaviour', *Geografiska Annaler* 74B: 189–97.

Sommerville, J. (1982) *The Rise and Fall of Childhood*, New York: Sage.

Sowden, S., Stea, D., Blades, M., Spencer, C. and Blaut, J.M. (1996) 'Mapping abilities of four-year-old children in York, England', *Journal of Geography* 95, 3:107–11.

Spencer, C., Blades, M. and Morsley, K. (1989) *The Child in the Physical Environment: The Development of Spatial Knowledge and Cognition*, Chichester: Wiley.

Stables, J. and Smith, F. (forthcoming) '"Caught in the Cinderella trap": narratives of disabled parents and young carers', in R. Butler and H. Parr (eds) *Geographies of Disability*, London: Routledge.

Stainton-Rogers, R. and Stainton-Rogers, W. (1992) *Stories of Childhood: Shifting Agendas of Child Concern*, Hemel Hempstead: Harvester Wheatsheaf.

Stea, D., Elguea, S. and Blaut, J.M. (1997) 'Development of spatial knowledge on a macroenvironmental level: a transcultural study of toddlers', *Revista InterAmericana de Psicologia* 31, 1: 141–7.

Steedman C, (1990) *Childhood, Culture and Class in Britain*, London: Virago.

—— (1995) *Strange Dislocations: Childhood and the Idea of Human Interiority 1780–1930*, London: Virago.

Stephens, S. (1997) 'Nationalism, nuclear policy and children in Cold War America', *Childhood: A Global Journal of Child Research* 4, 1: 103–23.

Valentine, G. (1996a) 'Angels and devils: moral landscapes of childhood', *Environment and Planning D: Society and Space* 14: 581–99.

—— (1996b) 'Children should be seen and not heard: the production and transgression of adults' public space', *Urban Geography* 17: 205–20.

—— (1997a) '"Oh yes I can." "Oh no you can't.": children and parents' understandings of kids' competence to negotiate public space safely', *Antipode* 29: 65–89.

—— (1997b). '"My son's a bit dizzy" "My wife's a bit soft": gender, children and cultures of parenting', *Gender, Place and Culture* 4: 37–62.

—— (1997c) 'A safe place to grow up? Parenting, perceptions of children's safety and the rural idyll', *Journal of Rural Studies* 13, 2: 137–48.

—— (1999) 'Being seen and heard? The ethical complexities of working with children and young people at home and at school', *Ethics, Place, and Environment* 2, 2: 141–55.

Valentine, G. and Holloway, S.L. (forthcoming) 'Virtual dangers?: geographies of parents' fears for children's safety in cyberspace', *Professional Geographer*.

Waksler, F. (ed.) (1991) *Studying the Social Worlds of Children: Sociological Readings*, London: Falmer Press.

Walvin, J. (1982) *A Child's World: A Social History of English Childhood 1800–1914*, Harmondsworth: Penguin.

Ward, C. (1978) *The Child in the City*, London: Architectural Press.

—— (1990) *The Child in the Country*, London: Bedford Square Press.

Ware, V. (1992) *Beyond the Pale: White Women, Racism and History*, London: Verso.

Weeks, J. (1985) *Sexuality and its Discontents: Meanings, Myths and Modern Sexualities*, London: Routledge.

Winchester, H.P.M. (1991) 'The geography of children', *Area* 23, 4: 357–60.

Winchester, H.P.M. and Costello, L. (1995) 'Living on the street: social organisation and gender relations of Australian street kids', *Environment and Planning D: Society and Space* 13, 3: 329–48.

Wood, D. (1985) *Outlook* 57: 3–20.

Wood, D. and Beck, R. (1990) 'Dos and don'ts: family rules, rooms and their relationships', *Children's Environments Quarterly* 7, 1: 2–14.

—— (1994) *The Home Rules*, Baltimore and London: Johns Hopkins Press.

Young, I.M. (1990) *Justice and the Politics of Difference*, Princeton, NJ: Princeton University Press.

Part 1

PLAYING

2

MELTING GEOGRAPHY

Purity, disorder, childhood and space

Owain Jones

Introduction

In this chapter I explore some aspects of the coming together of children's and adults' geographies in a particular place. My interest is in how adults' geographies present strongly striated space(s)[1] and how children's geographies take form, the best they can, by suffusing through the fabric of adults' geographies. This work has developed from Philo's (1992: 198) call to consider 'the possibility of uncovering something of the world inhabited by rural children, particularly the geographies of these worlds as structured "from without" and as experienced from "within"', but in this instance I hope the issues raised are of relevance to considering the encounter between child and adult geographies more generally.

This chapter makes connection with recent theoretical concerns with the social constructions of childhood (for recent summaries see Gittens 1998; James, Jenks and Prout 1998) and with how children as 'other' remain deeply mysterious presences in the midst of such adult constructions. How childhood is constructed by adults and the implications of this has now received some notable attention (Aitken 1994; Sibley 1991, 1995a; Stainton-Rogers and Stainton-Rogers 1992; Valentine 1996, 1997), but so far the implications of children's 'otherness' have not been tackled in a sustained way within the social sciences generally or geography in particular, perhaps because of the genuine difficulty of doing so. The otherness of childhood is profound, as many of the symbolic orders which routinely but deeply structure adult life, such as time, money, property, sex, mortality and Euclidean space melt away as one tries to see the smoother, or perhaps differently striated spaces of childhood. Of course all these things have profound impacts on children's lives, particularly issues such as poverty and abuse, but how children understand and manage these fabrics of social (dis)order is profoundly other. As Frönes *et al.* state:

29

although adults themselves have to be constrained into social order, much as Durkeim argued, children offer living exemplars of the very margins of that order, its potential disruption and, in fact, its fragility. Children, on a momentary basis, exercise anarchistic tendencies. . . . They are dedicatedly unstable, systematically subversive and uncontained and all of these manifestations are managed, barely, under the rubric of creativity, self-expression, primitiveness, simplicity or even ignorance.

(Frönes *et al*. 1997: 260)

Here I want to consider encounters between children's and adults' geographies through the ways in which the children I have observed[2] use some spaces for play within a small 'idyllic' village in southwest England. From this I want to draw out what I feel may be some useful conceptualisations about childhood and space. Clearly there is a need for geography and sociology to differentiate between the experiences of children of differing age and gender, and a need to take account of the spatial, social, cultural and family contexts in which children find themselves (Aitken 1994; Gittens 1998; James, Jenks and Prout 1998), but in the necessarily short space allocated here, I raise issues which I feel have a broad relevance for the geographies of childhood, and particularly those in the late first stages and second stages of childhood. These issues revolve around the concept of pure space and what I am calling 'otherable space'. First I am interested in how constructions of spaces as pure or defiled (Sibley 1995b) interact with adult constructions of children as 'little angels' or 'little devils' (Valentine 1996; Warner 1994) in ways which affect children's everyday lives. My focus then moves to otherable space where I consider to what extent the dominant, striated fabrics of adults' geographies are, or can be, rendered flexible or porous enough for children to form their own geographies within them. This question is vital because the opportunity for children to create their own geographies, in other words, to spatialise their lives according to their own rather than adult agendas, at least to some extent, is seen as vital to their self-expression and their 'development', and it is also one of the aspects of Apollonian[3] childhood (see Chapter 1) which is now considered to be so much under threat. The focus of this chapter to some extent moves backward and forward between media representations of childhood and the reality of 'lived childhoods', and this I hope reflects how the 'real' and 'imaginative' are dialectically bound together in the ongoing (re)construction of both lived childhoods and popular (and academic) accounts of them.

In recent decades a growing commentary of crisis has emerged concerning the status, condition, and even 'the end' of childhood (Postman 1982). Wallace (1995: 286) claims this to be 'a moment of massive anxiety in the West about the capacities, safety and the status of children'. As Wallace indicates, concern over childhood is composed of a number of elements, particularly, significant disruptions in adult constructions of what childhood is, the blurring of the boundaries between childhood and adulthood, and concerns for the lived lives of contemporary children. Modern (Apollonian) childhood is understood to have suffered 'erosion' (Suransky 1982) through a whole range of factors, particularly the loss of innocence as a result of the increased exposure of children to sexual and other knowledges, the commodification of childhood, and the spatial and temporal confinement of childhood produced by fear of accident and violence befalling the unattended child (McNeish and Roberts 1995). Reports on the increased spatial limitations of childhood (Hugill 1998; Moore 1997), and of growing instances of children's mental health problems (Mental Health Foundation 1999), provide compelling support that concerns about childhood are justified, even if the more profound questions of adult constructions of childhood remain relatively undisturbed. The also relatively undisturbed assumption underpinning this work is that, whatever the nature of childhood and the otherness which comprises it, the spatial interactions between childhood and adult geographies – the ways in which children and adults can share space equitably, and the degree of spatial flexibility or rigidity children experience – are of the utmost importance.

A village – an adult geography

Like most other places, the village of 'Allswell' can be seen as a manifestation of adult geography in that it is predominately constructed and ordered (both materially and symbolically) on adult terms and scales. Situated in southwest England, it consists of a settlement of some 82 houses loosely clustered around two junctions of winding country lanes. It thus has a simple network of roads (most without pavements, which is relevant to the movement of children) and a number of cul-de-sacs, on which are distributed houses, gardens, the Hall and its surrounding park garden, the pub and skittle ally, church and churchyard, green lane, cricket pitch, village hall and car park (which used to be the village school and playground until 1970). The village hall, which makes a small terrace with the pub, looks onto a small 'green' where annual fêtes and so on take place. No shops remain. Four farm complexes situated within the village, which consist of traditional stone structures, have been converted for residential use, though three small more modern pockets of agricultural infrastructure still remain. This domestic landscape is surrounded and sporadically penetrated by agricultural land. Much of this is 'standard' lowland agriscape which is intensively farmed for arable or livestock purposes, but a few areas are less intensively farmed. These are areas of rough pasture, land now in the set-aside scheme, fields attached to houses in the village which are used for keeping horses or for

31

'recreational farming', and some areas of sloping stony ground – the village lying partly across a small yet steepish valley cut by a stream, which for a short stretch runs alongside the village high street. The fields are divided up by hedges in various states of upkeep, and where these make 'awkward' corners in terms of mechanical farming, some small patches have been left to go 'wild' or even planted with trees, and one is the site of a new pond. A network of footpaths and bridleways radiate from the village, some following the course of the stream out into the 'open countryside'.

Local estate agents say it is a desirable location. Most of the houses and gardens are well maintained and the boundaries between them, and between the domestic spaces and agrispaces around the village, are clearly defined. Other geographies, such as a designated conservation area, and the very restrictive permitted development designations, 'overlay' this material space.

Many (adult) networks radiate in and out of the village. Those who do paid work do so predominately in nearby urban areas, and most people have extensive social networks beyond the village. However, there is a strong sense of 'neo-community' with a whole range of (adult) groups and activities, many of which orbit around either the church, the cricket club, the village hall committee or the pub, but between which there is considerable multiple 'membership'. These geographies are of course not fixed, complete or coherent, but instead are partial, contested and disjointed. Differing imagined (aged and gendered) geographies of the urban, the rural, nature, childhood, parenthood and family are folded into the specifically configured material and symbolic actuality of the place.

It is within this (inadequately depicted) adult geography that children live their lives. Clearly the patterns of children's lives are closely interwoven with the patterns of those they live with, and should not been seen as sharply separated, but within all this children do construct their own imaginative and physical geographies, and it is these, and here particularly the latter, which I am concerned with. It is within this generally highly ordered adult territory that the 'dynamic trajectories' of children's becoming and milieu mapping (Deleuze 1998) happen. The extent to which these trajectories can develop or are confined, the extent to which they are disruptive and subversive, the extent to which they are sanctioned or censured, depends on the conditions of constructions of this adult space.

Pure space and childhood

First I will address the issue of pure space, for this has some significance in adults' structuring of children's lives within the village, and is perhaps a key issue in adult understanding of space and childhood interactions. The (British) countryside is often constructed as an idyll, and as an idyll for childhood. This is reflected in a formidable set of discourses which range through literature, the media and the preponderance of rural imagery in children's literature (Aitken 1994; Bunce 1994; Hunt 1995; Jones 1997). As Miller (1997) put it, 'as part of our national mythology, we hold the country to be a good thing for children'. Such views have emerged strongly in research concerning rural communities by Valentine (1997), Little and Austen (1996) and Bell (1994). Allswell also fits very well into this generalised construction of the rural.

There is, now, a high ratio of professional 'service class' (Cloke and Thrift 1990) families with young children living in the village, and a number of parents interviewed said that a key reason for choosing the village as a place to live was to give their children 'a country upbringing'. One of the key imaginative geographies of the village is this notion of it being an idyll for childhood. One father told me:

> Well I think it's a nice environment, it's relatively quiet, safe, in terms of traffic and things like that kind. Pleasant community. So I think it has idyllic prospects for childhood . . . and you can watch things grow and play in the stream.

Other parents expressed similar views, often using the urban as a negative comparison,

> The children do appreciate waking up in the morning with the birds singing and no constant noise of cars going by every second. The space really, the space to go out and enjoy what you can see around you in the trees and flowers. The [lack of] pollution of course is a big factor. Of course because of the traffic thing I feel it is much safer here. I feel as well that they are very much part of the community which it is possible to have in a village rather than in a town.

At annual village events 'the village children' take symbolic centre stage, dancing around the May Pole at the May Day Fête, and being organised into a series of performances at the Village Social.

The construction of the rural, and specific rural places, as idyll(ic), are instances of the construction of space as pure (Sibley 1988, 1995b). Most forms of idyll, according to Eisenberg (1998: 143), are ways 'of denying or declawing change'. The reasons for this must in part lie in the observation that:

> familiarity and predictability are important for many people. There is a common desire to live in a place which is stable and orderly, where

social interaction entails what George Herbert Mead called 'a conversation of gestures', gestures which are mutually understood.

(Sibley 1999: 115)

The rural as idyll in general, and particular places which articulate this notion in some form of imagined actuality, represent a very powerful and familiar conversation of gestures. Bahktin (1981) makes the crucial point that pastoral idylls need a temporal dimension of stability and purity, and that children can become a central image within this. Sibley's (1999) concern with constructions of space as pure and impure, or even abject, revolves around the exclusions, oppressions and persecutions of others which might crystallise in and around these spatialities. My concern focuses more on how childhood interacts with these constructions (an issue that Sibley also addresses), not only in terms of the contested nature of the interaction of childhood and pure space, but also in terms of the more positive implications for childhood within pure space, which revolve around some children being given a degree of freedom because of the construction of purity. I hasten to add that questions of exclusions (and excluded children), and the often deeply evil and tragic means by which 'purity' is constructed, are paramount and that work is needed which reveals, and perhaps redefines what 'purity' is and what is its value.

Innocence is perhaps the key element of Apollonian constructions of childhood which dominate in many modern cultures. Childhood is seen as 'idyllic, carefree, happy' (Gittens 1998: 7). This legacy of romantic sensibilities also associates children with the natural and valorises children and nature as wellsprings for (re)purifying and invigorating the tarnished soul of modern (industrialised, urban) society. Perhaps inevitably the countryside has come to be seen as a childhood idyll because it is where the innocence of childhood can connect with innocent spaces of nature (Nahban and Trimble 1992).

The pure space that is 'Allswell' is in essence an adult geography, but it enables, to some extent, some (but not all) children to reconstruct this adult space in their own terms, because some parents will allow them the freedom to do so. Somewhat paradoxically the purity of the village acts as a form of control, which in turn enables some of the children some degree of limited freedom. Most parents told me that although their children have far less spatial/temporal freedom than they had had as children, they still felt that in the village they were prepared to allow the children a certain degree of autonomy (e.g. walking to a friend's house, leaving the domestic space of house and garden for play), which they would not if they were living in a differing, usually urban, context. One parent told me, 'I prefer the countryside [to bring up children], you have lots of places to play, the fields and whatever, compared to living in a city'. Some children have limited autonomy to exploit certain spaces. For example, at one end of the village the stream is followed by a footpath, and in one place is overhung by trees where the valley is quite steep and this makes what feels like a private, secretive space. Various cohorts of children have used this place, known as 'the den', for meeting up, and a base for activities. This was marked on a number of the maps drawn by the children, and two friends (girls aged 11 and 12 interviewed together) told me the den was 'somewhere you can go and sit and talk away from everyone else . . . everyone's in the house, it's really cramped, so you go to the den'. Another space used by successive cohorts of village children is one of the remaining farm yards with its two barns, which was also depicted in the children's maps.

Children of course can create their own geographies in spaces which are constructed as impure by some adults. Perhaps the most vivid imaginings of this are children on 'inner city estates'. Here children are often depicted in the media as being too free, to the extent that the state considers curfews, and asks parents 'Do you know where you child is right now?' Here freedom, usually a celebrated aspect of childhood, is seen as dangerous, and the children as feral, even vermin,[4] playing strange, dangerous games such as 'lift surfing' (*Guardian*, 30 May 1997), and reverting to *Lord of the Flies* savagery (*Daily Mail*, 27 October 1993). Clearly there are views which contest these constructions of urban space(s), particularly those of children and adults within them, and writers such as Philip Ridley (1995) who is trying to re-enchant the urban as a setting for children's stories. But as Scott, Jackson and Backett-Millburn point out:

> one crucial aspect of the spatial distribution of risk anxiety [of parents about children] is the difference between urban and rural locations. There are both material and imaginary differences between the city and the country . . . for example, the idea of the city as dangerous spaces haunted by the spectres of crime and violence versus the romanticised and nostalgic views of the countryside.
>
> (Scott *et al.* 1998: 700)

One mother I interviewed, who had moved to the village from an inner city area with her husband and two children, told me 'well, you see, he [son] couldn't be a wild thing in Crompton Road [their old address] without people telling him off and whatever, whereas out here [in the village] he can, can't he'. Another mother also talked about the effect the constructions of the village had on how children's behaviour was received – 'they can't do wild things in the city, can they, without, without sort of damaging things. Jack running around with a huge stick (here) sort of, it looks funny rather than menacing, doesn't it'. There is a need to unravel these interconnecting constructions of pure and defiled space, children as 'little angels' or 'little devils', the rural and the urban, and

35

the implications they have for the lives of children, but for now I want to consider the issue of otherable space.

Otherable space(s)

Reconfiguring space

As Allswell is constructed as a pure space, or to be more precise, as a significant number of parents/adults construct some parts of it as pure space, this gives some children limited opportunities to exploit some of the spaces of the village on their own terms. This makes it possible, through ongoing observation, to build up a picture of the interaction between the embryonic geographies which these children construct and the adult configured space in which they do this.

The ways that children use the space(s) show that a number of factors affect how children's geographies can spread through adult striated spaces. This occurs in all manner of quite subtle interconnecting ways. For example, children have a propensity to establish short-cuts and 'other' route-ways which disregard the established adult route-ways and boundaries of the village (see also Ward 1990). As Moore (1986: 56, 57) suggests in his analysis of children's 'flowing terrain[s]', 'their patterns of interaction [with a given terrain] are more intimate, fluid and intense' than those of adults. In addition to this, (younger) children seem much less clearly aware of the presence and/or significance of sharply defined boundaries of ownership, and private and public space. Consequently they may wander (or race) from road to garden, garden to garden, house to garden to house, garden to farmyard, garden to field, garden to lane to church yard, moving through the striated geographies of adult symbolic and material boundaries. The different experiences and practices of children vary considerably through the differing extents to which parents sanction this sort of movement, and stress issues of the recognition of other people's property, and also through the receptiveness, or perceived receptiveness, of the adult overseers of the differing spaces. Notions of childhood/country idyll put pressure on parents/adults to look upon this sort of movement more benignly, but this is intermixed with notions of fear for children (what might befall them), and fear of children (what they might 'get up to').

If this represents an example of (adult) striated space becoming smooth through the physical and imaginary geographies of childhood, it also is space which is becoming striated in differing ways. As Deleuze and Guattari (1988) stress, these relations are never fixed, but are in a constant ebb and flow of construction and deconstruction. The space(s) of the village will in fact be differently striated for children, in terms of other demarcations such as boring, scary, interesting, more or less likely to get told off, but the significant thing is that children do have some opportunity to operate their own spatialisations rather than remain utterly confined within the patterns of adults' geographies.

Adult spaces can be in some way 'otherable' in that children can use and reconstruct them without incurring the outright hostility and opposition of adults. First there are spaces of disorder where, for some reason or another, adult geography has abandoned or disregards the internal dynamics of a particular space, some of which then have the potential to become children's places. Second, there are what I call polymorphic spaces. These are spaces which, unlike monomorphic spaces, can somehow accommodate the differing uses of adult and childhood spatial configurations. Third, I consider how the variability and manipulability of a space adds to the potential for children to reconfigure adult space on their own terms. Fourth, I consider the degree to which boundaries which embed the adult geographies of the village are permeable. The more permeable the adult boundaries are, the more accessible are these other spaces, and the wider space itself can be reconfigured by children. Lastly I consider how children are opportunists in the exploitation of space, which leads to the important perspective of 'freeing up' space for childhood, rather than trying to (over) determine it. My intention is not to depict wider space as a jigsaw or mosaic of micro spaces which are sharply and stably defined, but as a more blurred, shifting pattern which nevertheless has discernible spaces within it, but which fuse, divide, overlap and superimpose onto one another.

Disordered spaces

There has been an ongoing concern over the negative impact of strict (adult) order on childhood (Sibley 1981, 1988, 1995a; Wood and Beck 1990). Such concern has surfaced strongly in the work of those who feel that the increasingly ordered or tidied-up nature of the countryside mitigates against childhood (National Children's Play and Recreation Unit 1992; Philo 1992, 1997; Santaniello 1978; Shoard 1980; Ward 1990). Although these arguments deal with an important issue, I feel that to some extent they may over-emphasise the need or even ability of ordering forces to eradicate all pockets of disorder, particularly given children's often small-scale, fine-grained relationship to space. What perhaps is of more importance in issues of children using 'other' space in the countryside are the spatial/temporal restrictions imposed on them by anxious parents, and/or the changing nature and patterns of children's activities.

The affinity between childhood and disorder has long been the subject of study and narration. The Opies (1969: 15) in their famous survey of children's games and folklore observed that 'the peaks of a child's experience are . . . occasions when he escapes into places that are disused and overgrown and silent'. They go on to report how a most brutal and unequivocal process of disordering, the blitz on London, produced bomb sites which were ideal for children's play. This literal smoothing out of the striated space(s) of house and street opened up possibilities for children, as recalled in John Borman's film *Hope and Glory* and Grahame Greene's (1970) story 'The Destructors'. Massumi (1992) urges us to 'cherish derelict spaces' in the quest for becoming other; in such spaces the otherness of childhood may also find expression. Moore (1986) and Ward (1978) show how spaces of disorder in urban areas are appropriated by children. Derelict spaces are often pivotal in children's books – such the old barn in Crompton's William stories, and the territory in which Richard Jefferies' story 'Bevis', unfolds; and in memories of (country) childhood such as those of Dylan Thomas (1965) and Edward Thomas (1938) (see Jones 1997). In and around Allswell there are numerous small pockets of 'abandoned' space, which some children use. For example, there is a small stony scarp which runs along one side of the cricket field over which the tall straggly hedge overhangs, which makes a space where children explore and play, particularly in the summer when matches are being played.

Polymorphic spaces

These are spaces which are in use within adult structures but which can also accommodate subordinate 'other' uses. Given that the majority of the landscape is adult ordered, these are possibly more significant to children than the abandoned corners considered earlier, particularly as the spaces in most intimate connection with the houses of the village tend to be those which are kept in a higher degree of order. Polymorphic spaces need to be distinguished from monomorphic spaces, and particularly intensive monomorphic spaces. The latter are spaces which are dominated by a single use that excludes the possibility of other uses, be it by children or other potential (even non-human) users. This distinction of space is closely related to Sibley's (1992) strongly and weakly classified space, but differs in that polymorphic spaces can have strong, clear boundaries, whereas in Sibley's schema, with weakly classified spaces the boundaries themselves become more vague. Examples of monomorphic spaces in this instance are the roads, the space of intensive agriculture, gardens which are too 'precious' for children to run about in. These are (material/symbolic) spaces devoted to a single use which, particularly if it is intensive in some way, excludes others. For example, I was told by older residents that they could remember games being played in the road, but this (play)space has been closed off by the increased intensity of its dominant use. Polymorphic spaces, on the other hand, can sustain alternative uses by children even in the presence of the dominant use. Examples of such spaces which children use are the two barns at Manor

Farm, which are used (in part) for storing hay but in which various cohorts of children have played; the farm track and concrete yard around the barns which are designed for access to the barn but which are also an excellent play terrain cycle track, and so on; a stand of conifers which are still fulfilling their intended function of screening the modern farm buildings from the neighbouring Hall, but which are also the site of a tree house, and other ground-level den structures that children have constructed; and less intensively farmed areas, and even the cricket field, where access by children (for certain activities at least) is not seen as problematic.

In the example of the stand of conifers where the tree house and dens are, and which is accessed by a small bridge across a ditch which children have constructed, although this is still part of the adult geography of the place, it has become a separate and quite private place for children. There is no need for adults to access the space for its dominate function to continue, and the children's activities do not compromise that function.

Some spaces can switch from being monomorphic to polymorphic through the cyclical nature of dominant adult use. A seminal example of this are the fields around the village in which crops are grown. These are considered to be no-go areas for children while the crops are growing, but after they have been harvested, for a period of time these become spaces which children can use, and this is an occurrence deeply embedded in portrayals of country childhoods.

Variable and manipulable spaces

Variety in the environments which children use for play is now seen as critical for children's ability to be able to construct their own worlds in ways which are satisfying to them. Moore (1986: 234, my emphasis), in his study of children's use of, and the potential improvement of, urban spaces for children, claims that 'access and *diversity* emerge as the most important themes in childhood-environment policy'. This variety should be seen in terms of differing micro-spaces, scales, surfaces, forms, materials, spectacles and opportunities. Consequently such variety is both spatial and temporal, and often there is a complex linkage between the two, with spaces being internally diverse but also changing over time. The yard and barns at Manor Farm do provide such a varied and varying environment. To detail this systematically would be a massive task, because children have such a fine-grained relationship with material space. This is captured by Moore:

> [The children] hopped, climbed, balanced, skipped, rolled, swivelled
> and squeezed through, on, over, around and inside their surroundings
> – using ledges, posts, walls, curbs, banks, bollards, doorways, steps

and paving stones – their movement choreographed by the landscape, as their bodies responded to every opportunity.

(Moore 1986: 56)

In the situations I observed, a large puddle after rain, or ice after a frost, a pile of scalpings (a kind of 'sticky gravel') temporarily tipped on the yard before being used to fill in muddy gateways or for other farm maintenance, the presence of mud and dandelions, the coming into season of blackberries and conkers, were among the many events which the children engaged with. Children do seem to want to make and manipulate objects (thus the success of Lego and other construction kit toys), materials (as in 'mud pies' and the ubiquitous sandpit), and, as far as is possible, the environment in terms of remodelled space. As Sobel points out,

it is crucial for children to participate in world-making or world-shaping activities. Children need the opportunity to create and manipulate . . . The creation of these worlds from plastic materials . . . gives children the opportunity to organise a world and then find places within it to become themselves.

(Sobel 1990: 8)

Of course, this can often create a tension with adult constructions of what a space should represent. Jacobs and Jacobs (1980) 'found that adults tend to emphasise the benefit of safe, secure neighbourhood playgrounds whereas the key to children's satisfaction lay within them being given the opportunity to design and modify their own environments' (cited in Aitken 1994: 131).

Crucial to these activities is the presence of loose material in the environment which is both safe(ish), and not restricted from use due to adult needs or curfews. Some rural/country spaces quite clearly can offer sticks, stones, mud, grass and other vegetation as loose material (Shoard 1980). Within such analyses vegetation is seen as a key element. Moore (1989: 3), for example, presents evidence of 'the extent to which imaginative play and creative social integration can be supported by a highly manipulative environment having plants as its primary play material'. All this, and an additional palette of stuff like timber, string, wire, is available in and around Manor Farm yard.

Permeability of boundaries

Boundaries are critical in the structuring of children's lives. These can come in both physical and symbolic forms and often are constructions combining both to varying degrees. In physical terms, Allswell and its environs are divided into a number of private and public spaces which are defined by boundaries of one kind or another. Children essentially have to fit in with these patterns and structures and thus they live in a world largely structured by forces which mostly ignore them as meaningful priorities when that world is formed and reformed. The more rigid this structuring is, the more it will constrain children's worlds within it, and the degree of rigidity of such structures is determined in part by the extent to which the boundaries within it are permeable or impermeable. If these boundaries are to some extent permeable to children, they have a chance to build their own geographies, to reorder the space to their own desires and in effect create a dimension parallel to that of the adult space, which itself continues to function.

In Allswell the relative purity of the space does mean that some of the internal boundaries are permeable to children. Sibley (1995b), in his work on boundaries and exclusions, makes a qualified agreement with Sennett, who talked of (North American) suburbs as being pure spaces. In such work the boundary under consideration is largely symbolic and encloses the entire space/group, and identifies it from the 'outside'. Such strong 'perimeter' boundaries thus enclose relatively pure spaces, which allows internal boundaries, say, between households or private and public space, to be weakened, and this may be particularly important for children, whose independent mobility is often mostly on a small, 'internal' scale (think of the children rushing around on bikes in Spielberg's suburban-set film *ET*).

This notion of the permeability of boundaries is manifested in very immediate ways within Allswell. Some parents have deliberately constructed or altered garden fences so that their children can move from the garden to either a farm yard or field beyond which is owned by someone else and where there is no 'official' right of way. Others have erected very secure, difficult to cross, boundaries around their properties and do not expect their children to venture beyond them unless accompanied or sanctioned, and then only via 'officially' designated routes (the gate, the drive, roads and paths). The former position is more common, but within that there are degrees of difference between parents who let their children wander into a nearby field as long as they remain within sight, or go to a friend's house, to a few who are prepared to let their children go beyond this immediate realm of surveillance.

Children as opportunists

Providing that children have access to the flexibility of this 'other geography' which permeates the adult geographies in the ways I have outlined, be this in any local area, rural, urban or otherwise, children do have remarkable capacity for responding to shifting, unexpected, often fleeting opportunities for expression. This means that adult structuring of children's play opportunities should not try to over-prescribe or anticipate what children may do. The children's opportunistic exploitation of situations I observed was remarkably responsive, with a fantastical mixing of the material and the imaginary.

Some of the hay in the big barn at Manor Farm is collected and stored in the form of big rectangular bales. This bale format, in terms of size and weight, seems less 'child friendly' than small bales, in that they cannot be used to make dens and do not make suitable sized 'steps' in the stack for chasing and jumping games. The big bales are brought in from the fields on a trailer, loaded three layers high with four or so bales per layer. Because of their size and slightly distorted form, the bales often do not stack tightly together. On one occasion a trailer was left loaded with bales for a day or so in the yard prior to unloading. Here I found a small group of children who had discovered that they could crawl into a small gap at the back of the load and crawl along a tunnel formed by the gaps in between the bales, and emerge at either the front or side of the load, and then jump onto a landing pad they had assembled from some loose hay (Plate 2.1). This play opportunity only existed for the day the load was there. The children had spotted it, and exploited it and, given that other possibilities could be expected to arise, did not seem to regret its passing. On another occasion, a long length of baler twine was found in the barn by four children,

Plate 2.1 Children as opportunists: exploiting a fleeting chance to use a load of big bales for a game. © Owain Jones

and they then played games of rescue and climbing across the ditch that cuts across the corner of the yard. These are just two of many instances where I observed children in this space build their activity around some event, circumstance, or artefact, that they encountered. The background to this is of course that the children have the spatial freedom to access and exploit this space, and that the respective parents' construction of the space prompts them to sanction this access.

Conclusion

Clearly there are issues of child safety to be considered here, and at all times. This is something that the parents whose children use the spaces I have briefly described are constantly and anxiously negotiating. But the basic point is that if adult space is (seen as) pure, and is to some extent polymorphic, permeable and variable, children (if also seen as pure) may be able, to some extent, to build their own geographies and within that find things to do and expressions to create. It is essentially romantic notions of the countryside and childhood which lie at the heart of these configurations of potential adult/childhood geographies. Although romantic notions of childhood are yet another adult projection onto the otherness of children, which is driven by adult agendas, they have the effect of creating space for the otherness of children because they involve granting children some form of 'freedom' and, importantly, may loosen regimes of surveillance and curfews in particular circumstances. Children do not readily adopt the generic identity thrust upon them, but they can operate under the cover it may provide. Although this may not be an ideal formation of adult–children relations, it would seem preferable to relations which are essentially hostile or neglectful to children.

The constriction of children's access to free time/space is seen as a major element of the crisis developing in our understandings of romantic childhood; as the *Sunday Times* (5 August 1995) put it: 'These days our children are not so much free-range as battery-reared' (cited by McNeish and Roberts 1995: 3). The scarce, ghettoised, commodified, institutionalised, regulated, over-determined 'places of play' which seem for some to be the only option, are not considered to be an effective substitute for a geography which somehow is more open and multidimensional, where the wider environment remains as a 'childhood domain'.

Children mostly live their lives within the warp and weft of the striations of adult space. These material, symbolic and disciplinary structures are both incidental and deliberate in their relation to children. Children's geographies operate within these patterns. The question is the nature of the interaction between the two. If adults' geographies are intensive, rigid and powerfully embedded, there may be little chance for children to build their own geographies, but if adults' geographies can be more permeable, heterogeneous and tolerant of otherness, then those in society most celebrated for their bodily and mental spontaneity, creativity, exuberance and mobility, may have the chance to express this in the creation of their own geographies within the adult world which, it seems, is bound to continue to be the dominant ordering of space.

Acknowledgements

The paper stems from research funded by ESRC Postgraduate Training Award ROO429334176. I am grateful to all those who contributed to my research. Thanks to Paul Harrison for guidance towards sources and ideas on smooth and striated space. I am grateful for the insightful and detailed comments and suggestions provided by the editors. Any remaining or compounded inadequacies rest with the author alone.

Notes

1   The term striated space is taken from Deleuze and Guttari's (1988) differentiation between smooth and striated space. The latter is seen as sedentary, over-determined, Euclidean, hierarchical, orientated, (metrically) divisible and multiple work space. The former is nomadic, folded, non-hierarchical, unorientatable, 'nonmetric' free-action space. Neither should be seen as pure or fixed categories but rather as constantly ebbing and flowing contested constructions and deconstructions of space(s). But smooth space is favoured as a territory for 'becoming-other' (Massumi 1992), for affirmative embracing of events.

2   This chapter stems from research done for my Ph.D. in my home village where I live with my partner and two children. The research entailed conducting a series of interviews with parents in the village (there being a number of households with children in the village); interviews with older residents about changes to the village and their memories of their and others' childhood; and analysis of depictions of rural childhood in literature and the media. This was used to construct accounts of local and more general adult constructions of 'country childhoods'. A landscape survey was conducted, looking in detail at the places where children played and did not play, within and near the village, in an attempt to interrogate suggestions (Ward 1990; Shoard 1980) that changes to the countryside were restricting country children's play opportunities. A programme of research with the children was conducted, using (depending on age) interviews, giving children cameras to record their outdoor activities, getting children to map the village, and also a process of (participant) observation. As a parent living in the village where my children and those of neighbours often played in spaces near our home, I have had the extensive opportunity to observe and photograph children playing and using certain spaces. This process initially amounted to a somewhat systematised extension of what I would consider to be normal parental practice of 'keeping an eye' on our (quite young) children and friends as they played in spaces away from our house and garden, and on other

children who also used the space. I have now observed various combinations of children playing in spaces close to our house over a number of years, and have tried to be aware of how children move around the village more generally. The bulk of this chapter is derived from these latter observations, but I have included a few fragments of interviews conducted with parents where appropriate.

3 Jenks (1996) classifies two basic constructions of childhood in modernity: Apollonian, which deriving from romantic foundations, sees children as innocent and pure and to be nurtured, and Dionysan, which sees children as the bearer of original sin and creatures to be tamed and formed into civilised beings. The Apollonian view of childhood has dominated in many modern, Western societies over the past century or so and it is this form of childhood which is apparently under threat from eroding pressures and corrupting influences of modern life.

4 For example, 'Ratboy'. This, as reported in the *Independent* (9 October 1993), was the title given by a national tabloid newspaper to a 14-year-old boy who was arrested for burglary by police, after he had been found hiding out in a ventilation shaft in a block of flats on a high-rise housing estate. The story was seen as emblematic of the idea of feral, 'vermin' children, 'infesting' run-down urban areas of the UK.

References

Aitken, S. (1994) *Putting Children in Their Place*, Washington, DC: Association of American Geographers.

Bahktin, M.M. (1981) *The Diologic Imagination*, Austin: University of Texas.

Bell, M.M. (1994) *Childerley: Nature and Morality in a Country Village*, Chicago: University of Chicago Press.

Bunce, M. (1994) *The Countryside Ideal: Anglo-American Images of Landscape*, London: Routledge.

Cloke, P. and Thrift, N. (1990) 'Class and change in rural Britain', in T. Marsden, P. Lowe and S. Whatmore (eds) *Rural Restructuring*, London: David Fulton.

Deleuze, G. (1998) *Essays Critical and Clinical*, London: Verso.

Deleuze, G. and Guattari, F. (1988) *A Thousand Plateaus: Capitalism and Schizophrenia*, London: Athlone Press.

Eisenberg, E. (1998) *The Ecology of Eden: Humans, Nature and Human Nature*, London: Picador.

Frönes, I., Jenks, C., Rizzini, I. and Stephens, S. (1997) 'Childhood and social theory' (Editorial Introduction), *Childhood* 4, 3: 259–63.

Gittens, D. (1998) *The Child in Question*, Basingstoke: Macmillan Press.

Hugill, B. (1998) 'Minded out of their minds: children trapped inside', *Observer*, 29 March.

Hunt, P. (1995) *Children's Literature: An Illustrated History*, Oxford: Oxford University Press.

Jacobs, E. and Jacobs, B. (1980) 'Children as managers', *Ekistics* 282: 135–7.

James, A., Jenks, C. and Prout, A. (1998) *Theorising Childhood,* Oxford: Polity Press.

Jenks, C. (1996) *Childhood,* London: Routledge.

Jones, O. (1997) 'Little figures, big shadows, country childhood stories', in P. Cloke and J. Little (eds) *Contested Countryside Cultures*, London: Routledge.

Little, J. and Austin, P. (1996) 'Women and the rural idyll', *Journal of Rural Studies* 12, 2: 101–11.

Massumi, B. (1992) *A User's Guide to Capitalism and Schizophrenia. Derivations from Deleuze and Guttari*, Cambridge, MA: MIT Press.

McNeish, D. and Roberts, H. (1995) *Playing it Safe: Today's Children at Play*, Ilford: Barnardos.

Mental Health Foundation (1999) *The Big Picture: Promoting Children's and Young People's Mental Health*, London: Mental Health Foundation.

Miller, J. (1997) 'Country life is a killer', *Sunday Times*, News Review, 27 April, p. 3.

Moore, R.C. (1986) *Childhood's Domain: Play and Place in Child Development*, Beckenham: Croom Helm.

—— (1989) 'Plants as play props', *Children's Environments Quarterly* 6, 1: 3–6.

—— (1997) 'The need for nature: a childhood right', *Social Justice* 24, 3: 203–20.

Nabhan, P.G. and Trimble, S. (1992) *The Geography of Childhood: why children need wild places*, Boston: Beacon Press.

National Children's Play and Recreation Unit (1992) *Children Today, A National Overview: Developing Quality Play Services*, London: National Children's Play and Recreation Unit.

Opie, I. and Opie, E. (1969) *Children's Games in Streets and Playgrounds*, London: Oxford University Press.

Philo, C. (1992) 'Neglected rural geographies: a review', *Journal of Rural Studies* 8, 2: 193–207.

—— (1997) 'Of other rurals', in P. Cloke, P. and J. Little (eds) *Contested Countryside Cultures*, London: Routledge.

Postman, N. (1982) *The Disappearance of Childhood,* New York: Delacorte Press.

Ridley, P. (1995) In discussion with Michael Rosen on BBC Radio 4, *Treasure Islands*, 19 April.

Santianiello, J. (1978) 'Rural deprivation', in *The Country Child*, Lincoln: Centre for the Study of Rural Society.

Scott, S., Jackson, S. and Backett-Milburn, K. (1998) 'Swings and roundabouts: risk anxiety and the everyday worlds of children', *Sociology* 32, 4: 689–705.

Shoard, M. (1980) *The Theft of the Countryside,* London: Maurice Temple Smith.

Sibley, D. (1981) *Outsiders in Urban Society*, Oxford: Basil Blackwell.

—— (1988) 'Survey 13: purification of space', *Environment and Planning D: Society and Space* 6: 409–21.

—— (1991) 'Children's geographies: some problems of representation', *Area* 23: 269–70.

—— (1992) 'Outsiders in society and space', in K. Anderson and F. Gale (eds) *Inventing Places: Studies in Cultural Geography*, Melbourne: Longman Cheshire.

—— (1995a) 'Families and domestic routines: constructing the boundaries of childhood', in S. Pile and N. Thrift (eds) *Mapping the Subject: Geographies of Cultural Transformation*, London: Routledge.

—— (1995b) *Geographies of Exclusion*, London: Routledge.

—— (1999) 'Creating geographies of difference', in D. Massey, J. Allen and P. Sarre (eds) *Human Geography Today*, Cambridge: Polity Press.

Sobel, D. (1990) 'A place in the world: adult's memories of childhood's special places', *Children's Environment Quarterly* 7, 4: 5–12.

Stainton-Rogers, R. and Stainton-Rogers, W.S. (1992) *Stories of Childhood: Shifting Agendas of Child Concern*, Hemel Hempstead: Harvester-Wheatsheaf.

Suransky, V.P. (1982) *The Erosion of Childhood*, Chicago: University of Chicago Press.

Thomas, D. (1965) *Portrait of the Artist as a Young Dog*, London: J.M. Dent & Sons.

Thomas, E. (1938) *The Childhood of Edward Thomas: A Fragment of Autobiography*, London: Faber and Faber.

Valentine, G. (1996) 'Angels and devils: moral landscapes of childhood', *Environment and Planning D: Society and Space,* 14: 581–99.

—— (1997) 'A safe place to grow up? Parenting, perceptions of children's safety and the rural idyll', *Journal of Rural Studies* 13, 2: 137–48.

Wallace, J.-A. (1995) 'Technologies of "the child": towards a theory of the child-subject', *Textual Practice* 9, 2: 285–302.

Ward, C. (1978) *The Child in the City*, London: Architectural Press.

—— (1990) *The Child in the Country*, London: Bedford Square Press.

Warner, M. (1994) *Managing Monsters: Six Myths of Our Time*, London: Vintage.

Wood, D. and Beck, R. (1990) 'Do's and don'ts: family rules, rooms and their relationships', *Children's Environments Quarterly* 7, 1: 2–14.

3

CHILDREN'S STRATEGIES FOR CREATING PLAYSPACES

Negotiating independence in rural Bolivia

Samantha Punch

Introduction

Recent research in the new social studies of childhood recognises that children are competent social actors who play an active part in their social worlds (Caputo 1995; Mayall 1994; Waksler 1991, 1996). Yet there are still relatively few studies which document the ways in which children devise ways to counteract adult's power and control over their lives. My aim in this chapter is to consider how rural children in Bolivia actively negotiate ways to assert their autonomy despite being constrained by adults who enforce boundaries of time and space (Ennew 1994; A. James 1993; Sibley 1995). Different aspects of children's and young people's lives are considered, including the nature of their social relationships at home, at school, at work and in the community at large. The research took place in Churquiales, a rural community[1] in Tarija, which is in the south of Bolivia, above Argentina (see Map 3.1). A sample of eighteen households were visited regularly in order to carry out semi-participant observation and interview all the household members including children, young people, parents and grandparents. At the community school a variety of techniques were used mainly with thirty-seven children aged between 8 to 14 years, which included classroom observation, photographs, drawings (Punch and Baker 1997), diaries and worksheets (Punch 2000, in press).

In this chapter I outline some of the structural constraints on childhood including the unequal power relationships between adults and children, and the work demands imposed upon children in poor rural areas. Subsequently I focus on how children use the spaces of work and school to negotiate time for play, thus enhancing their spatial and temporal autonomy.

48

Map 3.1 Map of Bolivia showing the study region Tarija

Constraints on children's time

In poor rural households where production systems strive to meet the family's subsistence requirements, labour needs are high. All family members above five years of age are expected to contribute to the survival of the household. Children's

49

work can be vital for survival or can help maximise household productivity levels. It includes domestic work such as food preparation and sibling care, animal-related work including feeding the animals and taking them out to pasture, and agricultural work such as weeding, planting and harvesting. Hence, children's free time is limited by their household duties. Parents may occupy their time with errands and chores for the household, but children undertake many responsibilities of their own accord. Many children are only able to play freely at home, when their parents are out, or when all the household tasks have been completed. They also have to combine their work with school. Consequently, in rural Bolivia, children's activities at school and work leave them little spare time. Adults, whether parents or teachers, restrict children's free time by giving them tasks to do and by controlling their time both at home and at school.

Children are also constrained by unequal power relationships within the household and discipline from adults and elder siblings. Adult discipline of children in rural Bolivia is based on a system of punishments rather than rewards, which vary from mild discipline such as shouting or threats, to harsher discipline such as hitting or beating. There are many times when refusal to do a job, or attempts to avoid it, provoke a harsh reaction from an authoritative parent: *Me pegan con un rebenque*[2] ('They hit me with a whip' – Miltón, 10 years). The type of punishment which a parent used varied greatly according to the strictness of that particular parent and the perceived gravity of the wrong-doing. Most often, to gain an obedient response, it is enough for parents to raise their voices or threaten the child with a physical punishment: *Siempre amanezco que los voy a pegar para que hagan caso* ('I always threaten that I'm going to hit them so they do as they're told' – Dolores, mother). However, if this fails, the child may be hit, or sometimes beaten: *Dale un látigito para que aprenda* ('Hit them so they learn' – Dorotea, mother). Thus, children are restricted from acting autonomously in certain situations where the parental response to such independent behaviour is to tame them with a stick: *Mi mamá me pega con un palito* ('My mum hits me with a branch' – Sabina, 14 years).

Adults in Churquiales often use a well known local superstition called the *duende* (dwarf) as a control mechanism over children's time to persuade them to work more and play less. In particular, this tactic is used to encourage them to come straight home from school rather than stopping and playing along the way. There were several different beliefs among participants about who or what exactly *duendes* were. Some people said that they were the rebellious souls of children who had died without being christened. Others said that they were like a mini-devil. Some people thought that *duendes* existed because of girls who had had abortions. The general image of the *duende* was as a short but plump child-like figure who wore a large, broad-brimmed hat but no other clothes: *Ha sido un hombre gordo y tenía un sombrero grande* ('It was a fat man and he had a big hat' – Cira, 9 years) or *He visto un hombre con sombrero y era como un niño* ('I saw a man with a hat and he was like a boy' – Santos, 14 years). It was generally believed that the *duende* could be seen only by children, and that it appeared only to children who played too much and were easily distracted: *Dicen que aparece a los chiquis que juegan mucho con muñeca* ('They say it appears to children who play a lot with dolls' – Felicia, mother) or who played instead of working:

> *Los adultos cuentan de los duendes a los niños para hacerles da miedo. Cuando no quieren hacer algo, como una tarea, o para que no jueguen demasiado, cuentan del duende.* ('Adults tell stories of the *duendes* to children to make them scared. When they don't want to do something, like a task, or so that they don't play too much, they tell them that the *duende* will get them.')
>
> (Nicolás, grandparent)

Not all children took much notice of their warnings: *Yo no conozco el duende* ('I have never seen the *duende*' – Celestina, 12 years) and *Como no se ve, uno no cree* ('As you don't see it, you don't believe it' – Tomás, 13 years), but some children were worried by it: *Tengo miedo de los duendes para ir ha clases* ('I am scared of the *duendes* on my way to school' – Rosalía, 11 years). Many parents also believed that the *duende* existed and they made the most of children's fear of it as a way to encourage them to do as they are told:

> *Les digo a mis hijos que no juegan. La jugada es tentación, mejor es trabajar o hacer una cosa. Los haceres se pasan si están en la jugada.* ('I tell my children not to play. Play is temptation, it's better to work or do something. Things may happen when you're too wrapped up in play.')
>
> (Ignacio, parent)

Children's spatial autonomy

Spatially Churquiales is a relatively large community which contains five main zones (Map 3.2): the centre, El Mollar (south), Josepillo (southwest, hill area), Pedregosa (far west, mountain area) and La Toma (west, by the river). Households in one of the four zones which is not part of the centre of the community are approximately half an hour up to over an hour's walk from the community square. To cross from the furthest point on one side of the community to the other would take over two hours on foot. The main form of transport within the community was on foot, the other was horseback. Children indicated that they did not regularly use the whole space of their community. They tended

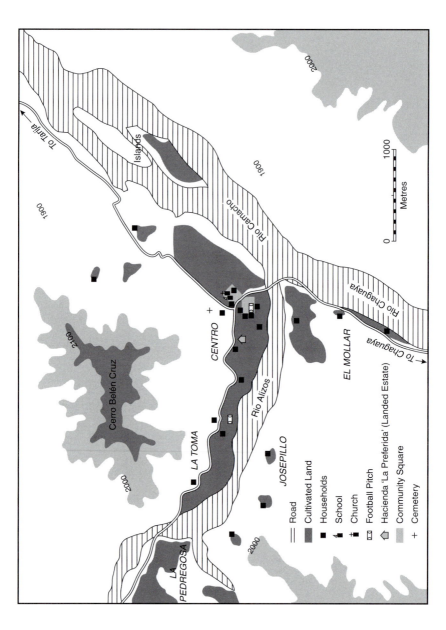

Map 3.2 A map of Churquiales indicating the five main zones of the community: *Centro, La Toma, La Pedregosa, Josepillo* and *El Mollar*

to use the area surrounding where they lived and the centre of the community. Children went regularly to the community square, as that was where the school was. They also used other facilities found in the square, including the football pitch, shops, the health post (*A aser curar al que esta enfermo* 'To make better whoever is ill' – Marianela, 10 years), and the church (*Para ir a escuchar la misa* 'To go and listen to mass' – Vicenta, 13 years). They frequently went to the square outside school hours to complete errands, or if a member of their household had been to town on the bus to shop, children were often sent to wait for that member's return to help carry the shopping back home.

Children went to the river or the irrigation canal nearest their homes on a daily basis to fetch water (except those who had recently acquired drinking water on tap in their homes), and also sometimes to bathe, to irrigate crops or to play. They also went up the hillsides nearest their homes to take goats and/or cows out to pasture, to fetch them in, or to search for a missing animal. Children used a wide range of spaces, often travelling great distances with animals or to carry out errands. Their mobility was closely linked to the demands of their household responsibilities and tasks. Adults considered children to be energetic and nimble, preferring to send them to do errands, or fetch and carry firewood and water. Therefore, children were usually more physically mobile and tended to travel greater distances each day than adults.

Such a finding contrasts with the daily territories of adults and children in the Minority World[3] where children are more restricted than adults because of fears for their personal safety. Children in urban societies have a limited use of space within the modern urban environment, which is considered to be threatening and dangerous to children, in terms of traffic, pollution and assault (Kovarík 1994; Valentine 1997). Consequently, children are confined to particular protected spaces which are usually controlled by adults: the school, the car or the house (Ennew 1994). They are rarely allowed to play freely in open urban spaces or out on the streets (Ward 1994).

In rural areas in Bolivia, children were not constrained by parental fears for their physical safety and they were free to roam the mountainsides and explore the surrounding countryside. In Churquiales, children's spatial experience of the local environment was therefore little different from that of adults. They had access to wide, open spaces and their use of space was more extensive compared with urban children. One family, who had recently migrated to the town of Tarija, reflected on their children's initial reactions to urban life:

Al principio los chiquis no se acostumbraron a quedar encerrados. Aquí es todo encerrado por miedo de las mobilidades. Ellos se quedaron tristes unos dos meses, echados, mirando debajo del portón. ('At the beginning, the children could not get used to being enclosed. Here it is enclosed because of our fear

of the cars. The children were sad for about two months, they used to
lie down and peer out under the gate.')

(Celia, parent)

The main parental fear for children's physical safety in rural Bolivia concerned
the danger of strong river currents during the wet season. Consequently, when
it rained, young children were not allowed to go to school if they had to cross
the river. In general, rural children experience substantial spatial autonomy (Katz
1993). In Churquiales, a child's daily movement to the square for school, to the
hillsides with animals and to the river to fetch water, was usually undertaken
alone, without seeking prior permission from parents. To travel further afield
within the community, such as to a zone furthest away from where they live,
most children would ask for permission to go. Such trips tended to be for specific
reasons such as to buy meat from someone who had recently killed a cow, or
A buscar un caballo ('To look for a horse' – Basilio, 11 years) or *Para visitar mi
tío* ('To visit my uncle' – Celestina, 12 years). On these occasions children
would usually be accompanied by a sibling or parent. However, not all children
told their parents where they were going. Sometimes they might disappear
without warning, or might go somewhere else instead. On their worksheets,
most of the older children (10–14 years) indicated that they had been to all, or
four, of the five zones of the community, but many of the younger children
(7–10 years) had only moved between the two or three zones nearest to their
home. Unsurprisingly, older children had greater freedom to move all around
the community and asked permission to do so less frequently than younger
children. Within the community there was little gender difference of the spaces
that children used, which supports the view that:

> Girls' freedom to move independently and to encounter directly diverse
> environments varies considerably across geographic settings, sometimes
> in ways that differ from commonly held impressions.
>
> (Katz and Monk 1993: 266)

Negotiating the use of time and space to create playspaces

The previous sections have outlined how children's use of time and space in the
community are structured by the demands of school or work – both activities
circumscribed by adults. In the Minority World many childhoods are associated
with school and play (Hunt and Frankenberg 1990). Play is a characteristic of
most childhoods (Ritchie and Koller 1964; Stone 1982), but in the Majority
World where many children work, the time which can be dedicated purely to

play is limited. In rural Bolivia, children's time was mainly taken up by school and work for their household, leaving them minimal spare time for themselves. This contrasts with the literature on children in the Minority World which argues that children's play is curtailed by the imposition of organised leisure (Ennew 1994).

Most research on play considers the meaning and consequences of play, particularly from a psychological perspective of child development (Cohen 1993; Herron and Sutton-Smith 1971; Pepler and Rubin 1982). Sociological studies of children's culture tend to focus on the types of play or the language which children use (A. James 1995; Opie and Opie 1982, 1991), but very few studies have considered the importance for children of combining play with other activities, such as work and school. In this section I show how children in rural Bolivia combine their use of time and space at work and at school in order to negotiate time to create playspaces for themselves.

When asked where they most liked to play, the children responded: on the football pitch and in the community square: both spaces at school (Plate 3.1). Nearly all their answers indicated spaces connected with going to school, because since children in Churquiales have many work responsibilities and live in households dispersed throughout the valley they have limited opportunities to meet other children apart from during school time. As Yolanda (aged 8) and Juana (aged 10) explained: *En otras partes no ay muchas chicas para jugar a las ollitas y muñequitas* ('In other places there aren't many girls to play kitchens and dolls

Plate 3.1 Girls playing football on the pitch in the community square. © Samantha Punch

with'), *Qui juego con mis amigos* ('Here I play with my friends'). Interestingly, parents, when asked the same question, said that they thought that their children liked to play at home. Some parents mentioned the football pitch, but only two parents said that their children most liked to play at school. Only one parent said that the children liked to play in the river, but several children referred to the river as one of their favourite places: *Para ir a pillar pescaditos* ('To go catching little fishes' – Valentina, 12 years) or *Para ir a vañarme* ('To go and bathe' – Oscar, 13 years). Therefore, parents did not necessarily know where their children most liked to play. This demonstrates children's ability to create their own playspaces away from parental control, of which parents might not be aware. It is also an example that indicates that parents' and children's views may differ considerably when contemplating the issues surrounding children's lives, since both have different perspectives, and children can be the most appropriate informants to consult about their own social worlds.

At school, play and having fun was not restricted to break time, but could also take place in the classroom, as well as by playing along the way to and from school. On the way to school children were less likely to play, since they were supposed to arrive in time for lessons. However, they often met their friends along the way, in which case their arrival at school might be delayed. By going to school alone, children experienced the freedom of physical independence rather than having to depend on being chaperoned. Children could be particularly resourceful when in search of fun and games. For example, on occasion, some of the pupils would meet on the way, and stop to play on the grass football pitch which was about twenty minutes' walk from the school. They would arrive late in class with excuses of having had to carry out chores at home. Sometimes they would never even go to school, but would stay and play by the river between their homes and the school. They would wait for the other children to return and would join them on their way home, pretending to their parents that they had been in school all day.

Semi-participant observation revealed that another strategy which children used as an opportunity to play with friends was to arrive at school before lessons commenced and to play in the square until the bell was rung. Children who lived closest to the school would often go early on purpose to play for a while, and sometimes the boys in particular would arrange to meet each other fifteen minutes early to play a game of football. When the bell was rung at the start of school or after breaktime, the pupils initially continued playing. They took five or ten minutes to gather and enter their classrooms. Sometimes when they were particularly immersed in a game, the teacher had to go in search of them and tell them that lessons were about to start.

On the way home from school, I observed that children took particular advantage of their time together by playing along the way and delaying their return home. They would stop and play in the river when it was hot, or look for *chasquitas* (a plant with many long thin leaves which could be plaited and made to look like dolls' hair) when they were in season. The boys often stopped to play marbles, or football on the grass pitch. The girls would see who could find the prettiest plants or flowers. Consequently children nearly always returned later to their homes than really need be, and their parents often complained to them about this. They often arrived half an hour later, but sometimes they would appear as late as four or five in the afternoon, when their parents expected them home by two.

When children were not at school, their play was carried out either where they worked or at home. Occasionally children went to a neighbour's or a nearby relative's household in search of friends or cousins to play with, but such a custom was rare because of children's responsibilities which they perform for their households. Sunday was when this might occur, but it was by no means common practice. The spaces of their household and work were vast, including the patio area of their house, the surrounding land and hillsides, and the areas where they went to fetch water or firewood and to look after animals. The temperate climate and the work demands of the rural lifestyle meant that much of children's time was spent outdoors rather than confined to the house. Such physical freedom in their natural environment not only allowed children spatial autonomy (Katz 1993), but also enabled them to command more control over their use of time. Since their use of space was extensive and often away from the gaze of adults, children had more opportunities to self-determine what they did in such spaces and how much time they spent there. Rural children's use of time was more restricted than their use of space, but children's spatiality enhanced their physical independence from adults and allowed them to gain more temporal autonomy.

Since the length of time that adults allow children to dedicate to play is limited, children devise ways of extending that time by combining play with work. In the Majority World there is no clear dividing line to separate work and leisure, and in many instances the two overlap. Children, in particular, merge the two activities, so that their work is also fun and allows them the social freedom to play. Certain jobs in Churquiales could easily be combined with play, such as looking after animals. Most of the children described this as a potentially boring task, as they just had to accompany the animals out to pasture to make sure they did not wander, enter fields and destroy crops, or get lost. Consequently, the children – both boys and girls – often sang loudly while taking the animals out or rounding them up, as they found themselves in wide open spaces on their own. Sometimes they took something with them to occupy themselves, such as a doll or a truck to play with. Other jobs could also easily be integrated with play. Fetching water could include playing for a while in the river, and banging home-made drums to scare birds away from crops was more like having fun making music rather than a chore. The ways in which children combine work with play has frequently been overlooked in the literature on children's work. The work of young children tends to be perceived negatively as a burden for the child, yet here it has been shown how children fuse the boundaries of work and pleasure, as well as counter the time and space boundaries imposed by adults.

Children are versatile and often extend the limited time which adults allow them for a particular activity by playing at the same time, making the most of snatched moments and being prepared to risk punishment if caught. Since time for play is short, children need to be inventive to enhance their possibilities for asserting their social autonomy. Children know to what extent they can stretch adult rules by prolonging play; they know how long they are expected to take to walk home from school or to take animals out to pasture or to respond to the bell rung for lessons. Yet sometimes they feel that it is worth breaking the adult-imposed rules for a particular game or social activity, and subsequently facing the consequences. Sometimes they find a suitable excuse to account for the extra time spent, such as having lost an animal for a while, or having helped someone else with a particular task. Such excuses can be useful strategies to avoid punishment if they feel it may be given. Therefore, despite children having limited command over their daily use of time, they develop ways to control the amount of time they assign to their social world of play by manipulating the spaces they use for work and school.

Conclusion

Children in Churquiales negotiate ways to assert control over their social world, especially their use of time and space which is largely restricted by adults. They find strategies to prolong play, and combine it with both school and work in order to play when and where adults may be unaware of their actions. These findings contribute to a sparse literature which considers the overlap between different childhood activities of work, school and play. In rural Bolivia, children's household duties of work provide them with an opportunity to acquire more spatial and temporal autonomy. Children's work allows them greater physical movement within their community, which is relatively unrestrained and where they experience quite a wide range of physical freedom. Their mobility is rarely motivated by leisure reasons and usually fills more practical specific purposes of work rather than pleasure.

Children's use of space in rural Bolivia is therefore not very restricted and enhances their physical independence from adults. However, the time available for play is more limited and this encourages children to negotiate ways to make their own time for their social world of recreation. Spatiality rather than tempo-rality is the vital component of children's strategies to create their own play spaces. Children do have boundaries set by adults that limit their possibilities for freedom, yet within these constraints they assert their autonomy and play an active role in their social worlds. Children may be physically dependent on adults, yet socially they can actively assert their independence in the everyday activities of their lives.

Although this chapter is based on the daily lives of children in one rural community in Bolivia, it should not be forgotten that most of the world's chil-dren live in rural areas of the economically poorer countries of the Majority World. In a global context, the majority of children work and go to school rather than have a childhood dedicated purely to play and school as they do in the Minority World. The experiences of children integrating work, school and play as described in this chapter reflect the majority experiences of childhoods. Paradoxically, Majority World childhoods are considered to be deviant when mea-sured against the ideal image of Minority World childhoods, which are perceived as 'an idealised world of innocence and joy' (James and Prout 1990: 4) where children are 'obliged to be happy' (Ennew 1986: 18). This image of carefree childhoods is perceived as the ideal norm to which all childhoods should aspire.

It has been argued that the 1989 United Nations Convention on the Rights of the Child has introduced universal standards for children which were based on the middle-class model of Minority World childhood (Boyden 1990). Children who do not live up to such idealism are perceived as being deprived of child-hood (Save the Children 1995). Majority World children should not be labelled as having abnormal childhoods; instead, it should be recognised that Minority World children tend to experience more privileged, protected childhoods compared to most of the world's children. Childhoods of the Majority World where children combine work, school and play should be the barometer against which the minority experiences of play are measured. This could lead to a rethinking of ethnocentric childhood policies such as the UN Convention on the Rights of the Child, which tries to impose a minority view of play onto the majority of children (Article 31). If policies for children are to offer effective support, they need to redress this imbalance of the perception of childhoods in the Minority and Majority Worlds and be based on the realities of children's everyday lives.

Notes

1 The community has a population of 351, which is spread amongst 58 households. It is 55 km from the regional capital, which takes one hour in a private vehicle but between three to five hours on the local twice-weekly bus. Most of the families own two to three hectares of land, which they mainly use to cultivate potatoes, maize

and a selection of fruit and vegetables. They also tend to own a small amount of pigs, goats and chickens, as well as a few cows. Most of their agricultural and live-stock production is for family consumption, but any excesses are sold in local and regional markets. The community has a small main square, where there are three small shops, a church, a medical post and the village school, around which there is a cluster of households, and the other households are more dispersed throughout the valley, which can be up to an hour and a half's walk away from the village square. Therefore, travelling distances can be long and difficult between the community and the town as well as within the community itself.

2   Quotations have been extracted from interviews and task-based methods carried out with children. When children have written their comments on worksheets or in diaries, the spelling has been left in the original to reflect the regional flavour of the Spanish language.

3   Minority World refers to the 'First World' and Majority World refers to the 'Third World'. This is because the Majority World has the greatest proportion of the world's population and the largest land mass compared to the smaller size of the Minority World. The use of these terms recognises that people who live in the Minority World tend to experience more privileged lifestyles (access to more resources, higher standards of living etc.) compared with the majority of the world's population. In addition, present terms used to differentiate the economically richer and poorer regions of the world are either incorrect (East–West, North–South) or have negative connotations for the poorer countries by emphasising what they lack (since they are developing, less developed, low-income, undeveloped, always striving to be what the First World already is). The terms Minority and Majority World are the only ones to shift the balance so that the richer countries are described in terms of what they lack: in comparison they have less population and a smaller land mass. Despite recognising that such a term unduly homogenises the 'Majority', an innovative shift in the balance tends to cause the reader to pause on reading the terms Minority and Majority World, and reflect on the unequal relations between the two.

References

Boyden, J. (1990) 'A comparative perspective on the globalization of childhood', in A. James and A. Prout (eds) *Constructing and Reconstructing Childhood: Contemporary Issues in the Sociological Study of Childhood*, Basingstoke: Falmer Press.

Caputo, V. (1995) 'Anthropology's silent "others": a consideration of some conceptual and methodological issues for the study of youth and children's cultures', in V. Amit-Talai and H. Wulff (eds) *Youth Cultures: A Cross-Cultural Perspective*, London: Routledge.

Cohen, D. (1993) *The Development of Play*, London: Routledge.

Ennew, J. (1986) *The Sexual Exploitation of Children*, Cambridge: Polity Press.

—— (1994) 'Time for children or time for adults', in J. Qvortrup, M. Bardy, G. Sgritta and H. Wintersberger (eds) *Childhood Matters: Social Theory, Practice and Politics*, Aldershot: Avebury.

Herron, R. and Sutton-Smith, B. (eds) (1971) *Child's Play,* London: John Wiley & Sons.

Hunt, P. and Frankenberg, R. (1990) 'It's a small world: Disneyworld, the family and the multiple re-representations of American childhood', in A. James and A. Prout (eds) *Constructing and Reconstructing Childhood: Contemporary Issues in the Sociological Study of Childhood*, London: Falmer Press.

James, A. (1993) *Childhood Identities: Self and Social Relationships in the Experience of the Child*, Edinburgh: Edinburgh University Press.

—— (1995) 'Talking of children and youth: language, socialization and culture', in V. Amit-Talai and H. Wulff (eds) *Youth Cultures: A Cross-Cultural Perspective*, London: Routledge.

James, A. and Prout, A. (1990) *Constructing and Deconstructing Childhood: Contemporary Issues in the Sociological Study of Childhood*, London: Falmer Press.

James, S. (1990) 'Is there a "place" for children in geography?' *Area* 22, 3: 278–83.

Katz, C. (1993) 'Growing girls/circles: limits on the spaces of knowing in rural Sudan and US cities', in C. Katz and J. Monk (eds) *Full Circles: Geographies of Women over the Life Course*, London: Routledge.

Katz, C. and Monk, J. (1993) 'Making connections: space, place and the life course', in C. Katz and J. Monk (eds) *Full Circles: Geographies of Women over the Life Course*, London: Routledge.

Kovarík, J. (1994) 'The space and time of children at the interface of psychology and sociology', in J. Qvortrup, M. Bardy, G. Sgritta and H. Wintersberger (eds) *Childhood Matters: Social Theory, Practice and Politics*, Aldershot: Avebury.

Mayall, B. (ed.) (1994) *Children's Childhoods: Observed and Experienced*, London: Falmer Press.

Opie, I. and Opie, P. (1982) 'The lore and language of schoolchildren', in C. Jenks (ed.) *The Sociology of Childhood: Essential Readings*, London: Batsford Academic.

—— (1991) 'The cultures of children', in F. Waksler (ed.) *Studying the Social Worlds of Children: Sociological Readings*, London: Falmer Press.

Pepler, D. and Rubin, K. (eds) (1982) *The Play of Children: Current Theory and Research Contributions to Human Development*, Vol. 6, Basel: Karger.

Punch, S. (2000, in press) 'Multiple methods and research relations with children in rural Bolivia', in C. Dwyer and M. Limb (eds) *Qualitative Methodologies for Geographers*, London: Arnold.

Punch, S. and Baker, R. (1997) 'Visual representation: using drawings and photographs as research methods with children in Nepal and Bolivia', paper presented at Urban Childhood conference in Trondheim, June 1997.

Ritchie, O. and Koller, M. (1964) *Sociology of Childhood*, New York: Meredith.

Save the Children (1995) *Towards a Children's Agenda: New Challenges for Social Development*, London: Save the Children.

Sibley, D. (1991) 'Children's geographies: some problems of representation', *Area* 23, 3: 269–70.

—— (1995) *Geographies of Exclusion: Society and Difference in the West*, London: Routledge.

Stone, G. (1982) 'The play of little children', in C. Jenks (ed.) *The Sociology of Childhood: Essential Readings*, London: Batsford Academic.

Valentine, G. (1997) '"Oh yes I can." "Oh no you can't.": children and parents understandings of kids' competence to negotiate public space safely', *Antipode* 29, 1: 65–89.

Waksler, F. (ed.) (1991) *Studying the Social Worlds of Children: Sociological Readings*, London: Falmer Press.

—— (1996) *The Little Trials of Childhood and Children's Strategies for Dealing with Them*, London: Falmer Press.

Ward, C. (1994) 'Opportunities for childhoods in late twentieth century Britain', in B. Mayall (ed.) *Children's Childhoods: Observed and Experienced*, London: Falmer Press.

4

THE 'STREET AS THIRDSPACE'

Hugh Matthews, Melanie Limb and Mark Taylor

Introduction

In this chapter we consider the importance of the 'street' in the lives of a group of young people aged between 9 and 16, living in three edge-of-town council estates within the East Midlands. We use the term 'the street' as a metaphor for all public outdoor places in which children are found, such as roads, cul-de-sacs, alleyways, walkways, shopping areas, car parks, vacant plots and derelict sites. Our attention focuses on two issues. First, we examine a growing post-modern assumption that local 'streets' and neighbourhoods are of declining importance for young people's identities and lifestyles (Featherstone 1991). There are a number of reasons for such a view, but two popular conceptions prevail, both age-related. On the one hand, moral panics have constructed images of public space as dangerous and unsafe, where young children are vulnerable and under threat from the social and physical fabric of places (Cahill 1990; Valentine 1996a). This view projects young people as potential victims, under attack from unruly gangs, prone to the ravages of strangers and threatened by the excesses of environmental dangers, such as traffic, and desperately in need of protection and care. An alternative view relates to older children who are (re)defined as the problem. In this case, their visibility in public places is seen as discrepant and undesirable. Young people, here, are the polluting presence, because by congregating together they are seen to be challenging the hegemony of adult ownership of public space (Breitbart 1998; Katz 1998; Matthews, Limb and Taylor 1999; Sibley 1995). When 'read' together, these negative discourses account for a supposedly profound feature of contemporary life, that is, a progressive retreat from the 'street' by urban children. Whether 'angel' or 'devil' (Valentine 1996b), many young people are portrayed as having withdrawn into the haven of their homes, lured by the excitement of computer games (McNamee 1998). Qvortrup *et al.* (1994) suggest that, in effect, young people are being increasingly confined to acceptable 'islands' by adults and so are spatially outlawed from society. In this chapter we argue that although there is evidence for a general exodus of this kind (Valentine 1996a), this experience is not universal. For a substantial residual of young people, the 'street' remains an important part of their everyday lives, a place where they retain some autonomy over space. We give emphasis to the continued significance of the 'street' for young people so that their right of presence in public places is recognised.

Second, we consider whether the 'street' is the archetypal gendered environment that has commonly been depicted (Cohen 1972; Griffin 1985; Hall and Jefferson 1976; McRobbie 1991). Such a view considers outdoor places to be a principal domain of males, providing core venues for boys to play out their masculinities. In this scenario, girls are seemingly invisible, as if female street use is morally taboo, at best, something which is of little significance and so justifiably ignored and, at worst, something which concerns only an exceptional minority who are better written-off as troublesome others. Those studies which have identified girls as users of places tend to suggest that they gravitate towards either their own bedroom or those of a friend, or local shops and indoor malls, where they play out their femininities (Pearce 1996; Roman and Christian 1988). We will show that the 'street' provides an important social venue for many young girls and that their use of public outdoor spaces not only rivals that of boys, but also involves ways not frequently recognised in debates about gendered space.

In developing these two strands, we propose that there will be parts of the local neighbourhood which constitute important cultural spaces for young people. As such, some 'streets' become places which are identified as safe space, affording what Soja (1996) has termed 'thirdspace', where young people can gather to affirm their sense of difference and celebrate their feelings of belonging. In essence, these places are 'won out' from the fabric of adult society, but are always in constant threat of being reclaimed. Here we consider the 'street' not as an 'inert backdrop against which social practices unfold' (Pile and Keith 1997:101), but as 'lived space' (Soja 1997), a site for the development of cultural identity. In so doing, we attempt to connect the real (material geographies of place) with the imagined (symbolic geographies of space) to better understand how the social construction of identity is mapped and performed. Bhabha (1994: 219) refers to how the regulation and negotiation of spaces are continually remaking boundaries, 'exposing the limits of any claim to a singular or autonomous sign of difference – be it class, gender or race'. The same can be proposed for age and the articulation of cultural differences in space that are produced through generational circumstance. Central to Bhabha's thesis has been the understanding that identity is produced through (in)between spaces which provide 'the terrain for elaborating strategies of selfhood – singular or communal – that initiate new signs of identity, and innovative sites of collaboration, and contestation' (Bhabha 1994: 1). We suggest that young people on the 'street' can be likened to groups (in)between, 'neither One nor the Other, but something else besides, in-between' (Bhabha 1994: 224): that is, set between the freedom and autonomy of adulthood and the constraints and dependency of infancy, neither adult nor child, 'angel nor devil', situated in imagined communities (located in thirdspace).

We see a link here, too, with the ideas of Corsaro. His focus is on children's transitions to adulthood and childhood as social form. Corsaro challenges the traditional thinking about children and how they are socialised into becoming adults and, instead, introduces the concept of 'interpretive reproduction' (Corsaro 1997: 18). The notion is important in that it signifies that children are agents in their own development (see also James, Jenks and Prout 1998) and in addition agents in their own locality. The term 'interpretive' is an attempt to capture the innovative and creative aspects of children's participation in society and suggests that children participate in their own peer cultures by creatively taking or appropriating aspects of the adult world in order to address their own concerns. The term 'reproduction' proposes an idea that children are not passively internalising society, but are actively contributing to cultural production and change. Nonetheless, as others have recognised (Qvortrup et al. 1994) the term implies that children are also constrained by societal reproduction, that is, their lives are bound up with the societies, cultures and spaces of which they are members. In these senses, the 'street' represents a border zone, a place where young people may develop their own identities, but rarely overturn the hegemony (adults).

Study area and method

The study areas comprised three edge-of-town council estates in Corby, Kettering and Daventry. These estates form distinct morphological units, mostly separated and isolated from surrounding middle-class housing. Each of the areas is blighted by high unemployment and above-average levels of crime, and there are a large number of lone parent families, living in poor and often unsatisfactory conditions. A Young Person's Support Index,[1] a composite measure used to identify children and families in need by the Northamptonshire Social Services Department, placed each of these estates in the highest category of need within the county. The data upon which our observations are based were derived by means of a door-step questionnaire survey (n = 320), semi-structured interviews with groups of young people on the 'street' and in-depth discussions held over three sessions with three groups aged 10–11, 13–14, 15–16. Given the predominant social background of the three estates (less than 2 per cent from Asian and Afro-Caribbean backgrounds), our results largely describe the geography of white, working-class children. In all instances the nature of the project was fully explained to the young people, participation was cleared by letters of consent from parents/guardians and there was no coercion to take part. Each member of the research team was registered as an outreach youth worker. Wherever possible, we incorporate the voices of young people into our text in order to get as close as possible to their lifeworlds.

The first part of our discussion considers who is out and about on the 'street', the nature of their activities, the reasons they give for being there and the meaning that the 'street' has for them. We then focus on the lifestyles of two groups of young girls distinguished by those who mostly stay at home and those who use the 'street' daily to meet friends.

Living on the edge: out and about on the streets

Within these three areas young people were highly visible on the 'streets'. There was a great deal of reliance on the local neighbourhood as a social venue. Some 74 per cent of those interviewed claimed that they hang around with friends on the 'street' on at least one occasion per week during the school summer holidays, with 43 per cent meeting on the 'street' on more than five days a week. There is some variation by sex. Only 18 per cent of boys never used the 'street' in this way, compared to 34 per cent of girls. Yet, some 40 per cent of girls regularly used the 'street' as a place to meet up with friends (five or more days per week), a percentage only slightly exceeded by the boys (47 per cent). When asked to categorise themselves as either an indoor or outdoor person, 85 per cent of boys claimed to be the latter. None the less, a significant 62 per cent of girls saw themselves in the same way. These results not only draw attention to how the 'street' is an important social venue for some groups of young people, but also impress how outdoor places are not spaces for boys alone.

> I go out every night of the week . . . I just wander the streets . . . the Grange mainly . . . About five of us . . . mostly girls but it depends where we are.
>
> (13-year-old girl)

> I have to stay in once a week and do all the chores and everything, homework and then I go out everyday of the week.
>
> (14-year-old girl)

For many, urban spaces are places of social inclusion, arenas where young people can get together to share and enjoy a range of informal activities, unhindered by the adult gaze (Plate 4.1). In so doing, girls and boys often use the 'street' in different ways. For example, the main activity reported by girls was talking and chatting with friends (46 per cent), whereas boys are more likely to see the 'street'

Plate 4.1 Hanging around with friends: developing 'self', developing identity.
© H. Matthews, M. Limb and M. Taylor

as a venue for informal sports, such as football, skateboarding and rollerblading (50 per cent). Like Lieberg (1995), we note that for both boys and girls there is a strong sense of theatre and of being on display when out and about.

> (*What is it about going out onto the streets that's better than staying in?*) Somewhere to hang about. And you meet all your mates and you get a social life. And you'll know they'll be there . . . so you can go anytime and you know they'll be there.
>
> (13 year old girl)

> It's where everybody comes . . . to meet . . . It's where everybody hangs out, sits on walls, smokes cigarettes, chats. It's . . . a place where you're likely to meet up with lots of other people . . . Where you meet your mates . . . try to figure out what our next move is going to be.
>
> (16-year-old boy)

> You can have a laugh round here. You can just relax. Everyone's down to earth . . . You don't have to be better than everyone and you get to sit down and talk and you're not pathetic if you've got a different point of view to everyone else. Everyone's just equal down here. It's chill.
>
> (14-year-old girl)

Some people find youth clubs and the Community Centre a bit dodgy . . . Everyone's just standing there and you are always being watched . . . And you can't do anything . . . but its a laugh when you go out to the parks and stuff . . . cause you meet all your mates.

(14-year-old girl)

Just to see you mates and that.

(14-year-old boy)

Yeah and catch-up with gossip. It's like if you stay at home you think 'oh, what am I missing?' And then you go to school next day and find out and wish I had been out.

(13-year-old girl)

Being together also confers a feeling of safety. Some 61 per cent of girls and 39 per cent of boys suggested that they were afraid of the 'street' when out alone, though when with friends less than 32 per cent of girls and only 19 per cent of boys expressed the same fear. Coping strategies of this kind not only dissipate feelings of apprehension about being out alone, but also encourage a broader range of place use and at different times (for example, after dark). Similar findings are reported by Pearce (1996) in her study of teenage girls in East London and by Watt and Stenson (1998) in their work within 'Thamestown'.

(*Do you feel that it's safe here?*) Yeah. If we're in a group. I always feel safe with these lot . . . cos I know that they won't suddenly turn on you for no reason.

(14-year-old girl)

It's close to home. Loads of people live round here so it's not that far to travel . . . There's a chip shop here. We can buy food and drink. Get to see all your friends. You can always go to someone's house. There's always somebody around here just in case there's trouble.

(10-year-old girl)

'Streets' are also places of social inclusion, not only in respect of being places without adults but also in terms of providing delimited geographies of social belonging through gang membership. In his seminal study of working-class youth in Sunderland, Corrigan (1979) highlights how 'street activity' was always carried out in a group. In our survey, about a third of the boys and girls interviewed said that they were members of a local gang. Evocative names are used to distinguish one group from another and each gang's 'patch' is marked out on the mental landscape.

> We're just a posse really. Just a load of friends. We always stick together . . . This is our territory . . . no townies allowed . . . no Southbrook . . . It's so exciting here.
>
> (14-year-old girl)

> We don't have group fights. I mean if one person in this group has an argument, we always try to sort it out . . . All the groups that sit round, they usually by the end of the night escalate down this end . . . they always come over to us cos we're mates with the group over there and over there.
>
> (13-year-old girl)

> . . . you get like different areas, like the Westies . . . you have a Westies crew and a Pleasure crew at Pleasure Park and a North Park crew . . . The Pleasure Park crew, they normally meet on a Friday night . . . we just have a laugh there . . . there's a lot like us. (*So the different crews from different places don't mix with each other?*) Sometimes it depends whether they have a fall out. (*What sort of age group is this?*) They go from about 12 to 18 and some people who's like losers come down until about 21. (*Is this mixed groups?*) Yeah.
>
> (13-year-old girl)

> . . . sometimes when we are going to play football or something . . . there's this gang of teenagers and they just won't let you play . . . they just shout at us . . . 'you can't come over here cause this is our bit, cause we're here'.
>
> (10-year-old girl)

'Streets' are places of meaning to many young people. They afford spaces where social conventions can be contested and independence asserted. In this sense, 'streets' are places where adultist conventions (constraints) and moralities about what it is to be a child, that is, less-than-adult, can be put aside. The result is that for a number of young people 'streets' become spaces between cultures, sites that are temporarily outside of (adult) society. From this perspective 'streets' are fluid domains, not 'a dualistic territory of transparent same and invisible other' (Rose 1995: 369), but instead a thirdspace set between the same (adult) and other (child). As such, they comprise dynamic zones of tension, 'discontinuity' and 'disjuncture', an 'interstitial space' (Bhabha 1994: 219), where young people can express feelings of belonging and of being apart and celebrate a developing sense of selfhood. We suggest that 'streets' can therefore be grouped among those places where the 'newness of hybrid identities (can) be articulated' (Rose 1995: 369).

> At my age I don't need looking after, I can fend for myself now, even though I do have to be in by 9 o'clock . . . You have to get used to the streets cos you're going to be on your own one day anyway. (*girl A*). Yeah . . . So you've got to get used to it . . . So why not learn now . . . Independence. (*girl B*)
>
> (13-year-old girls)

> In a house you have to be quiet cause if you're too loud then your Mum tells you to get out or Dad. But then if you are in the park you can do what you want, shout your head off, run around. (*boy A*) You can be really loud and talk about really rude things, can't you? (*girl C*) Yeah, but in your house you can't, can you? (*girl D*) No cause . . . No cause your Mum and Dad might hear you . . . and you get well embarrassed if they find out things that you talk about . . . And you really don't want them to know. (*girl C*)
>
> (14-year-old boys and girls)

Although hanging around on the street confers a certain social credibility, there is a strong sense amongst young people that the major reason for being there is that they have nowhere else to go and nothing much else to do. Katz (1998) acknowledges that, increasingly, young people are faced with lessening choice and fewer opportunities of where they can go, without adult interference. She describes an eroding ecology of youth and childhood, an outcome of the 'pernicious effects created by the decay and outright elimination of public environments for outdoor play or "hanging out"' (Katz 1998: 135).

> (*What do you normally do around here?*) Sit outside the shops. Have a laugh. (*Why here?*) Because there's nothing else to do. There's not much places else to go. There's nowhere else out . . . Yeah, other places you go you get moved by the police all the time.
>
> (14-year-old boy)

> There's nowhere else to go . . . There's only the base (youth club) on a Friday and there's nowhere else to go. (*boy A*) There's the chip shop . . . and the pub. We would be allowed in the pool room there . . .

sit with our drinks, non-alcoholic, but there's a woman in there who chucks everybody out. You got to be over 16. (*boy B*)

(13- and 14-year-old boys)

There's nothing to do in Daventry . . . play football, that's it. Nothing really. (*Why do you hang around here?*). Nothing else to do (*chorus of voices*). Nowhere else to go . . . It's where all the action is.

(mixed group aged between 13 and 17)

Can't meet in parents' houses . . . Some of our parents don't really like it and there's a lot of us as well. (*girl B*) Youth club will ruin your street cred . . . Ruin you whole style. It's just not cool. (*girl A*)

(13-year-old girls)

Yet for many young people the 'street' is the only place where they can meet up with friends. Lieberg (1995) points out that youth, unlike adults, have little access to 'backstage space'. Adults can withdraw to different places connected with work, membership and residence; in many cases young people have no opportunity, access or obvious right to such places. In this sense, the 'street' acts as a marginal space for young people, a place they occupy by default, as they lack the power to control other places. Thus, for many the home is often perceived to be an unsatisfying social environment. This is adult space in which children are often denied privacy, and 'boundary' disputes are common (Sibley 1995; Valentine, Skelton and Chambers 1998). In our survey, 54 per cent of the sample rarely (35 per cent) or never (19 per cent) had friends visit them at home, a feature more pronounced for boys (57 per cent) than for girls (41 per cent). Conversely only 18 per cent were allowed to have friends at home on a daily basis (boys, 13 per cent; girls, 23 per cent).

My mum moves me on from my house. (*13-year-old girl A*) Yeah I get moved on from my house. (*boy A*) Yeah I do . . . so do I (*boy B and girl A*) I get chucked out of the house about seven. (*So your Mum doesn't like you staying in, in the evenings?*) No . . . cause I don't know what to do and I just annoy her. (*boy A*) Yeah, same here . . . I annoy my Mum and then she starts screaming and then I get chucked out of the house . . . I just get in the way, get out and play. (*girl C*) My parents send me out because I don't do anything. (*girl A*)

(mixed group aged 13 and 14)

When taken together, these results emphasise the importance of outdoor places in the lives of these young people and suggest that the retreat from the 'street' is likely to be a socio-spatially selective phenomenon. For the working-class boys and girls of these three estates, the 'street' was an important cultural space. Here young people hang out and develop some of their closest connections and this is where social identities are constructed. From this perspective the 'street' affords venues where young people can stand apart, not necessarily in gestures of resistance, a point we shall develop later, but simply to assert their independence and to get away from the adult gaze. Through their daily and immediate contact with their neighbourhoods, some 'streets' become stamping grounds, places of safety, where young people can congregate with little threat of intervention by others. These 'streets' are the thirdspace of urban youth.

Living on the edge: indoor and outdoor girls

In order to explore further the importance of the 'street' in the lives of young girls, we focus on two groups distinguished by their levels of 'street' use. One group comprises those girls who hang around with friends on five or more days per week on the 'street' (n = 64; 40 per cent of sample) and for means of comparison, we identify a second group, girls who never or only occasionally (less than once per week) meet their friends on the 'street' (n = 74; 46 per cent of sample).

Although at the age of 9, twice as many girls are likely to be with friends indoors rather than on the 'street', with respect to age composition there is no statistical difference between the two groups.[2] From the age of 10 onwards there is no clear pattern of distinction, which suggests that age is not a sole determinant in who is out and about on the streets (Table 4.1). It would seem that despite growing public concern about stranger-danger and fear of assault, within these three estates girls of all ages are hanging out in public spaces.

Table 4.1   Girls hanging around on the 'street': by age and frequency

Age group	< 1 day/week*	> 5 days/week
9	13	6
10	7	8
11	12	7
12	2	4
13	13	10
14	7	11
15	5	10
16	15	8
Total	74	64

*Frequency of meeting friends on the 'street'

There is a strong temporal dimension to hanging out (Table 4.2). Those girls who go out only rarely to meet friends are required to be back home at an earlier time than their counterparts.[3] For example, 79 per cent of these girls are back home by 10 p.m. on a summer's evening, compared to only 59 per cent of those who regularly use the 'streets'. Yet, up until this time the majority of both groups of girls (57 per cent) are still out and most will have been on the 'street' for three to four hours. Far from being an insignificant presence in space and time, girls are habitual users of the 'street'.

> We're out here from about half six to about half nine, ten. (*13-year-old girl*) And during the day as well when we're off school. At the end of the day maybe we go in for our dinner, but sometimes we just stay out all day.
>
> (14-year-old girl)

> I am waiting for a new curfew, it's going to be in my birthday card, I bet it's going to be something like 10 o'clock . . . Boys get longer curfews . . . there's a load of year seven boys out later than me, little year eights and they're out at like half past ten and I'm in my bedroom.
>
> (13-year-old girl)

Hanging around is a diverse activity. For those girls who are regularly out with friends, time is spent doing many things. There may be little that can be described as spectacular, dirty or dangerous, the absence of which may be one reason why girls were once taken as 'invisible' in discourse about youth

Table 4.2   Times that girls are required to be back at home during summer holidays: by frequency

Time back home (pm)	< 1 day/week*	> 5 days/week
5	5	1
6	5	1
7	4	4
8	11	7
9	13	5
10	16	15
11	8	17
12	6	6
Anytime	6	8
Total	74	64

*Frequency of meeting friends on the 'street'

subcultures (Valentine *et al.* 1998), but less than 2 per cent replied that they were doing nothing in particular. Activities ranged across 'just talking' (69 per cent) to 'playing' (14 per cent), 'informal sports and leisure pursuits' (11 per cent) and shopping (2 per cent).

From the surveys, too, there was little sense that the 'street' was a symbolic site for cultural resistance. Girls were out and about because the outdoors comprised a set of places where they could get on with the ordinary and do what they wanted to do. Like Lieberg's (1995) teenagers in Sweden, the young girls who formed part of the survey were attracted to public spaces in order to be with others of their own age and because the outdoors conferred places where special things happened. The 'street' was not a domain where adult values were challenged as a matter of course. Corrigan (1979), too, notes that 'rule breaking acts' are not the normality of hanging out. However, adult control was often present and both groups of girls were aware and responsive to parentally imposed spatial limits. For example, although caretaking practice in these neighbourhoods was formed in such a way that girls were enabled to hang around within their neighbourhoods, 79 per cent of those who were infrequent 'street' users and 70 per cent of 'street' regulars reported clearly delimited place bans. Of those girls experiencing such a ban, only a minority in each case (17 per cent and 32 per cent respectively) were prepared to contest these fixed geographies and most considered these limits to be fair and reasonable (81 per cent and 77 per cent respectively).

None the less, in carrying out the mundane and usual, collisions and confrontations with adults formed an almost daily part of 'street' life for these girls. Contestations of this kind can be likened to a symbolic staining of places by the hybrid (Bhabha 1994), acts through which the prescribed duality of the powerful (adult) and the less powerful (child) become blurred. Being with friends when outside the home is very important to most young people. Yet it is when young people congregate together that they are often seen as 'out-of-place' and their behaviour construed as threatening. For example, Sibley (1995) notes how a group of teenagers hanging around the swings and slides of a public park in the early evening are commonly seen as discrepant, simply because their non-conforming use of place challenges adult expectations. Our survey highlights a complex 'turf politics', whereby the geography of hanging out is, on the one hand, both choreographed and conducted by sets of deeply invested and widely shared (adult) cultural values and, on the other hand, representative of an attempt to usurp place by children.

(*What do the people round this area think of you hanging around?*) They hate us. Some of them moan all of the time . . . We've had (name) mum calling us slags when we just sit here and talk as friends. We've got people . . . who are calling us tarts cos we are sitting with . . . It's like everyone judges us on the way we look just because they think we're louts and layabouts . . . we just sit down here.

(14-year-old girl)

. . . say you were standing round, hanging round there, people report you to the police and say, oh they're breaking into houses and you get blamed for things you didn't do.

(13-year-old girl)

(*Has anybody ever told you to go away?*) From North Way . . . They come out and say get out of here . . . Old biddies who live there . . . They say no ball games and there's not a sign anywhere.

(12-year-old girl)

(*Has anybody ever tried to stop you hanging round?*) Yeah . . . W and D up at the pub always come out and say f——— off. (*What reason did they give?*) Because it's (the pub) right by the side of the launderette where it's nice and warm for us to go and sit when it's really cold . . . and they don't like it. They can be as noisy as they want in the pub but not noisy as we want outside.

(11-year-old girl)

People from the houses by the park said we were making too much noise when we was in the park . . . They says if you're going to make that much noise you can go some place else . . . There were some older kids over the fields and they dropped like pop bottles and crisp packets . . . and we got the blame for it. She said if you leave that much mess again then she was going to get the council.

(10-year-old girl)

I think they're narrow-minded because they forget they were children and they forget that they had to go somewhere.

(13-year-old girl)

What is evident from the comparison of these two groups of girls is that those who are regularly on the 'street' differ very little from their more home-based counterparts, either in the way that they use places when out or in respect to their regard for adult caretaking values. In essence, the 'street girls' are not the 'she-devils' of urban folklore (Griffin 1993; McRobbie 1991). Denied opportunities at home and with very little else to do locally, for many young girls in these edge-of-town estates, the 'street' provides the only venue for social interaction, a place where they could spend most of their leisure time, in mixed groups, with little fear or threat.

Conclusion

In this chapter we have sought to contest two popular conceptions of urban living, that outdoor places no longer matter to urban children and that 'streets' are the domain of boys, providing only sites of indifference to girls. We will discuss each of these points in turn.

In contrast to those studies which have suggested that there has been a general retreat indoors by urban children, within the harsh and blighted neighbourhood settings afforded by these three estates, many young people, including younger children (those aged 9–11 years), relied heavily on outdoor spaces during their free time. Contrary to common media imaging, where the 'street' is seen as dangerous and threatening, and adult stereotyping, where young people are seen as better out of public places, these neighbourhoods provided security, freedom and social opportunity. 'Streets' comprised (semi)autonomous space or the 'stage' where young people were able to play out their social life, largely unfettered by adults. As Corrigan (1979) and Valentine (1996b) have suggested, for many young people public space becomes reconstructed as their own private space, especially with the retreat of adults after dark. Many of the children in our studies were drawn onto the 'streets' because there was a strong feeling that there was nothing else to do and because their parents did not want them at home. We recognise that this is not the case everywhere, nor are all young people inevitably drawn into becoming habitual 'street'-users. For as Valentine (1996a: 587) comments, 'the experience of childhood has never been universal'. What we suggest is that there are multiple realities and therefore multiple child-hoods, layered, on the one hand, by such contingencies as place (e.g. social and environmental opportunities within the neighbourhood and the home), parental caretaking practices (e.g. notions of what makes a 'good parent', peer group pressure, children-in-care) and the socio-personal characteristics of parent and child (e.g. age, sex, social class, ethnicity, educational background, income, ill health, disability); and on the other hand, by the 'agency' of children themselves (e.g. personality, lifestyle, choice).

What we have also attempted to show in this chapter is that the 'street' is not a homogeneously gendered space, a place reserved for boys to explore their masculinity. Instead, the 'street' is a key domain for many young girls. Rather than being unimportant or undesirable occupants of public space, girls use the outdoors in a variety of ways. For some girls, the 'street' is their principal meeting place, where they can hang around with friends, chat and wait for things to happen. These are girls who are not 'out of control', nor are they stepping out with the sole intent to challenge adult values. Rather, they represent a group, who through their own volition and with a certain amount of parental encouragement, have (re)defined the 'street' as a 'their' space. However, McRobbie (1991) notes the continuing 'invisibility' of girls in debates about public space, as if their presence there is morally reprehensible. Griffin (1993: 128), too, laments 'the patronizing complacency of traditional "malestream" youth research' and the tendency to 'romanticize the "macho" sexism and racism of the lads'. In their recent review of feminist contributions and critiques to youth studies, Valentine *et al.* (1998: 17) suggest that in the late 1990s some change is underway 'and that there is now a much broader consideration of what young women do and what constitutes the "distinctive elements" of their culture' (see, for example, Blackman 1998; Dwyer 1998; Leonard 1998; McNamee 1998). None the less, there is still work to be done in order to render the position of young girls more visible upon the urban landscape.

Lastly, in this chapter we have considered how the 'street' may be interpreted as thirdspace. In a recent review essay, Rose (1995: 372) is critical of the 'incorporeal' in Bhabha's conceptualisation; for her, 'his spatiality remains analytic, not lived'. We suggest that by focusing on the everyday experiences of young people within their neighbourhoods there is considerable strength in such a metaphorical concept, particularly as it challenges the duality of difference between adult and child. Crang (1998: 175) speculates that 'it may be that cultures are not holistic ways of life, but are constructed by people from the assemblage and reassemblage of the social fragments that surround them'. We propose that the 'street' provides a space for such (re)construction. These are places where young people can piece together their own identities, celebrate an emotional sense of togetherness and stand apart, if only temporarily, from the adult world which surrounds them. 'Streets' are spaces betwixt and between cultures, neither entirely 'owned' by young people nor fixed as adult domains. As such, they comprise 'contradictory cultural landscapes' (Crang 1998: 175) from which signs of autonomy and separateness are both created and inevitably blurred.

Acknowledgements

We are grateful to all the young people who took part in our survey. Thanks, too, to the youth workers of Northamptonshire Youth and Residential Service, Gemma Hartop who assisted in the data collection and Barry Percy-Smith for the excellent photograph. The project was funded by a grant from the ESRC 'Children 5–16: growing into the 21st century research programme' (Award No. L129251031).

Notes

1 The index was based on a set of indicators from the 1991 Census associated with the risk of children being taken into care: single parent family; > 4 children in family; Afro-Caribbean ethnic group; tenure; overcrowded households; children aged 10–17; unemployment rate.
2 Kolmogorov–Smirnov two-tailed test, $X^2 = 1.48$, df = 2, not significant at $p = 0.05$.
3 Kolmogorov–Smirnov two-tailed test, $X^2 = 7.3$, df = 2, $p = 0.05$.

References

Bhabha, H. (1994) *The Location of Culture*, London: Routledge.

Blackman, S. (1998) 'The school: "poxy cupid!" An ethnographic and feminist account of a resistant female youth culture: the new wave girls', in T. Skelton and G. Valentine (eds) *Cool Places: Geographies of Youth Cultures*, London: Routledge.

Breitbart, M. (1998) 'Dana's mystical tunnel: young people's designs for survival and change in the city', in T. Skelton and G. Valentine (eds) *Cool Places: Geographies of Youth Cultures*, London: Routledge.

Cahill, S. (1990) 'Childhood and public life: reaffirming biographical divisions', *Social Problems*, 37: 390–402.

Cohen, S. (1972) *Folk Devils and Moral Panics: The Creation of Mods and Rockers*, London: Basil Blackwell.

Corrigan, P. (1979) *Schooling the Smash Street Kids*, Basingstoke: Macmillan.

Corsaro, W. (1997) *The Sociology of Childhood*, Thousand Oaks, California: Pine Forge Press.

Crang, P. (1998) *Cultural Geography*, London: Routledge.

Dwyer, C. (1998) 'Contested identities: challenging dominant representations of young British Muslim women', in T. Skelton and G. Valentine (eds) *Cool Places: Geographies of Youth Cultures*, London: Routledge.

Featherstone, M. (1991) *Consumer Culture and Postmodernism*, London: Sage.

Griffin, C. (1985) *Typical Girls? Young Women from School to Job Market*, London: Routledge and Kegan Paul.

Griffin, C. (1993) *Representations of Youth*, Cambridge: Polity Press.

Hall, S. and Jefferson, T. (eds) (1976) *Resistance Through Rituals*, London: Unwin Hyman.

James, A., Jenks, C. and Prout, A. (1998) *Theorizing Childhood*, Cambridge: Polity Press.

Katz, C. (1998) 'Disintegrating developments: global economic restructuring and the eroding ecologies of youth', in T. Skelton and G. Valentine (eds) *Cool Places: Geographies of Youth Cultures*, London: Routledge.

Leonard, M. (1998) 'Paper planes: travelling the new grrrl geographies', in T. Skelton and G. Valentine (eds) *Cool Places: Geographies of Youth Cultures*, London: Routledge.

Lieberg, M. (1995) 'Teenagers and public space', *Communication Research*, 22, 3: 720–44.

Matthews, H., Limb, M. and Taylor, M. (1999) 'Reclaiming the streets: the discourse of curfew', *Environment and Planning A* 31, 10: 1713–30.

McNamee, S. (1998) 'The home: youth, gender and video games: power and control in the home', in T. Skelton and G. Valentine (eds) *Cool Places: Geographies of Youth Cultures*, London: Routledge.

McRobbie, A. (1991) *Feminism and Youth Culture*, Basingstoke: Macmillan.

Pearce, J. (1996) 'Urban youth cultures: gender and spatial forms', *Youth and Policy*, 52: 1–11.

Pile, S. and Keith, M. (eds) (1997) *Geographies of Resistance*, London: Routledge.

Qvortrup, J., Bardy, M., Sgritta, G. and Wintersberger, H. (eds) (1994) *Childhood Matters: Social Theory, Practice and Politics*, Aldershot: Avebury.

Roman, L. and Christian, L. (1988) *Becoming Feminine: The Politics of Popular Culture*, London: Falmer Press.

Rose, G. (1995) 'The interstitial perspective: a review essay on Homi Bhabha's The Location of Culture', *Environment and Planning D: Society and Space* 13: 365–73.

Sibley, D. (1995) *Geographies of Exclusion*, London: Routledge.

Soja, E. (1996) *Thirdspace: Journeys to Los Angeles and Other Real-and-Imagined Places*, Oxford: Blackwell.

—— (1997) 'Planning in/for postmodernity', in G. Benko and U. Strohmayer (eds) *Space and Social Theory*, Oxford: Blackwell.

Valentine, G. (1996a) '"Oh yes I can." "Oh no you can't.": Children's and parents' understanding of kids' competence to negotiate public space safely', *Antipode* 29, 1: 65–89.

—— (1996b) 'Angels and devils: moral landscapes of childhood', *Environment and Planning D: Society and Space* 14: 581–99.

Valentine, G., Skelton, T. and Chambers, D. (1998) 'Cool places. An introduction to youth and youth cultures', in T. Skelton and G. Valentine (eds) *Cool Places: Geographies of Youth Cultures*, London: Routledge.

Watt, P. and Stenson, K. (1998) 'The street: it's a bit dodgy around there', in T. Skelton and G. Valentine (eds) *Cool Places: Geographies of Youth Cultures*, London: Routledge.

5

'NOTHING TO DO, NOWHERE TO GO?'

Teenage girls and 'public' space in the Rhondda Valleys, South Wales

Tracey Skelton

Introduction: Where are the girls?

To date there is very little material within geography studies which focuses on the experiences of teenagers (but see Matthews, Limb and Taylor 1999). Girls, who are placed in that in-between age of not really being children anymore and not yet adult women, have also been a neglected group within existing studies of 'youth'. Where teenage girls have been taken into account, it has been through their involvement in home spaces and within domestic relations. What girls, and especially white working-class girls, do outside has not been part of empirical focus (however see Chapter 4).

In this chapter I investigate the ways in which a small group of teenage girls living in the Rhondda Valleys, South Wales, UK use the streets as places of leisure and how these uses intersect with their regular participation in the youth clubs and activities of the Penygraig Community Project.

Initially I consider the role that binaries play in Western thought, in particular the construction of the adult/child binary and the ambiguous position teenagers occupy within it. I then examine the role of the public/private binary and explore the ways in which the girls experience and construct their own leisure experiences. In many ways the girls can be conceptualised as occupying an ambiguous and often uncomfortable position of being the 'wrong' age, being the 'wrong' gender and being in the 'wrong' place. However, as I illustrate, the girls resist such a marginal positioning, especially through their friendships.

Binary understandings of space and childhood

Binaries in Western thought

Western cultural meanings and interpretations tend to rely on the construction of binaries, and one side of the binary is invariably seen as inferior, secondary, or problematic in relation to the other (Sibley 1995a; Women and Geography Study Group 1997). Such fundamental binaries which influence much of our thinking, our perceptions and the ways in which we are socialised include the following: culture/nature, mind/body, white/black, public/private, work/home, male/female, boy/girl, adult/child. In each case the former is designated as having greater importance, power, prestige or status. In some cases the binaries are interconnected. Male/female can be overlapped with public/private, the male being associated with the first space, and the female with the latter. Similarly, adults may be associated with consciousness, rationality and reason, hence the 'mind' is seen to predominate in their decision-making and the ways in which they conduct their lives. Children, on the other hand, are understood to predominately respond to 'bodily' needs, the immediacy of now, and are unable to respond to a rational argument about patience, appropriate times and schedules (Aitken and Herman 1997).

Binaries tend to imply discrete and separate spaces or identities, and as such they can act as boundaries (Sibley 1995a, 1995b). 'The public/private distinction works by establishing a boundary between the public and private sphere and by drawing a line between public and private space' (Laurie et al. 1997: 114). Within the social sciences the vast majority of research has focused upon the public sphere, in particular on the workplace and related institutions. The binaries of public/private, male/female and adult/child have been pervasive in the literature on teenagers and in many ways much of the earlier research served to reinforce common understandings of these so-called differences and separations (Furlong and Cartmel 1997; Griffin 1985; McRobbie and Garber 1976).

The adult/child binary

If we focus on the adult/child binary, one of the immediate problems is how we define those two sides of the divide. When is a person a child and when is a person an adult? In terms of social research such a distinction may be problematic, but for legislation and a whole range of social institutions (schools, the courts, pubs) and social professions (social workers, teachers, youth workers) age is prescriptive and based upon the date of birth and so lived through the body in terms of years. The social and cultural liminal spaces occupied by teenagers are elsewhere firmly scripted into what these older children/younger adults can or cannot do (James 1986; Valentine, Skelton and Chambers 1998).

The girls I worked with were aged 14 to 16. Hence they were at an ambiguous age, at once children (in full-time compulsory education), teenagers (socially defined as difficult, moody, rebellious and trouble-making), and young people (celebrated as the future, full of energy and life). Holstein-Beck argues that while there are several ways in which to define the teenage years, a common way is defining it as a transitional period between childhood and adulthood:

> the teenage years come after childhood and are influenced and shaped in varying degrees by childhood experiences; at the same time, the young person's anticipated future also affects his or her teenage experiences and decisions.
>
> (Holstein-Beck 1995: 100)

This ambiguity of meaning is not just about an academic definition, but is something experienced by teenagers themselves and through social and cultural perceptions of them. Throughout the small group discussions I had with the girls, they would fluctuate between mature, reasoned arguments and fits of giggles. Sometimes I was interviewing wise young women and other times I was caught up in silly jokes and girlish stories. This slippage of behaviour is apparently part of teenage girls' behaviour in places where they feel comfortable and at ease with their friends. In her discussion of teenage girls and their construction of style through fashion, Hillevi Ganetz talks about the girls when they go home to their (bed)rooms together in order to view and play around with the clothes they have purchased. She says that the 'girls' room is a protected space where girls can be serious, giggly, childish and adult – without adults' supervision or control' (1995: 88). The importance of spaces and places where teenage girls can perform aspects of their identities and play around with meanings of femininity is a significant focus of this chapter, although here the emphasis is on teenage girls within the so-called 'public' space.

There are other serious issues at play around the ambiguous positioning of teenagers. One of the other facets of Western conceptual frameworks is the uncomfortableness and disquiet around the inbetweeness of something or someone who does not lie neatly on one side of the binary or the other. We see this in public and private reactions to transgendered people (neither male or female), the problems of defining and legislating for marginal spaces (doorways where the homeless sleep, roadsides where peace protesters pitch their camps), and the anxieties and moral panics around teenagers (not children and not adults). Such moral panics have a continued and profound impact on the social relations surrounding children and young people (Boëthius 1995; Lucas 1998; Valentine 1996b). Within geography, David Sibley's work on exclusion examines the messiness of teenagers as a social group:

[The] child/adult illustrates a . . . contested boundary. The limits of the category 'child' vary between cultures and have changed considerably through history within western, capitalist societies. The boundary separating child and adult is a decidedly fuzzy one. Adolescence is an ambiguous zone within which the child/adult boundary can be variously located according to who is doing the categorising. Thus, adolescents are denied access to the adult world, but they attempt to distance themselves from the world of the child. At the same time they retain some links with childhood. Adolescents may appear threatening to adults because they transgress the adult/child boundary and appear discrepant in 'adult' spaces . . . These problems encountered by teenagers demonstrate that the act of drawing the line in the construction of discrete categories interrupts what is naturally continuous. It is by definition an arbitrary act and thus may be seen as unjust by those who suffer the consequences of the division.

(Sibley 1995a: 34–5)

Early work on children and young people largely focused on boys, in particular their activities in the public spaces of the street (Corrigan 1979). Girls were assumed to be located within the private sphere, at home, a supposition which continues to be repeated in contemporary texts. In their 1997 essay on leisure and lifestyles, Andy Furlong and Fred Cartmel (1997) recognise that the past two decades have seen an increase in the amount of free time young people have and that the range of leisure pursuits open to them is greater than ever before. However, they stress that studies of young people's leisure indicate that there are strong gender distinctions in activities and these are more significant than those associated with class or race.

It has been noted that girls are less leisure-active than boys; they are expected to spend more of their free time helping out in the home and are required to return home earlier in the evening. Moreover, they tend to have fewer resources to help enrich their leisure time; they receive lower wages and less pocket money and have higher 'self-maintenance' costs.

(Furlong and Cartmel 1997: 55)

Consequently, even in texts of the late 1990s the interpretation of research is that girls are firmly located within the domestic, the private sphere. However, recent research on children's use of computer and video games suggests that public space may no longer be a key arena for boys, but that the home is becoming a significant site of male leisure (McNamee 1998). This reinforces

Mica Nava's (1992) assertion that more attention needs to be paid to the ways in which gender relations and power are played out in the domestic space.

As we will see below, for the teenage girls I worked with this inbetweeness dictated by their age impacts upon their everynight[1] choices of where they can go and what they can do. In addition, this is compounded by their gender, their class position and their geography.

The public/private binary

The social and cultural adult/child binary is given spatial significance when mapped onto the public/private binary. There are assumptions about the spaces which adults and children are supposed to inhabit. This is further complicated by social differentiation based on gender: there are often different 'rules' about uses of space and the timing of those uses for girls and boys.

The public/private binary is heavily gendered, but it is also heavily aged. Through her work with parents and teachers, Gill Valentine (1996a, 1996b) describes how parents of young children fear that older teenage children constitute a threat to them within public space, whereas parents of teenagers worry about the risks which they might be exposed to on the street, including drink, drugs and violence. These dangers are firmly imagined in the public space through adult anxieties, and the desire to protect the young. Valentine (1996b: 590) argues that 'public space becomes defined as "naturally" and "normally" an adult space'. She demonstrates that in order to protect their spatial hegemony, adults develop a range of strategies to restrict children's access to public space, including reporting children to the police, interrogating children on the street, and calling through the media or democratic processes, for more controls such as curfews and increased use of CCTV cameras. She concludes that adults act upon an

> assumption that the streets belong to adults and children should only be permitted into public spaces when they have been socialised into appropriate 'adult' ways of behaving and of using space. Autonomous young people appear to be automatically perceived to disrupt the moral order of the street.
>
> (Valentine 1996b: 596)

As we begin to deconstruct this binary of public/private, not only is there recognition of the blurring of the boundary of such a division, but there is a deep questioning of the meanings of the terms themselves. If we accept the commonplace notion that 'public space' is a sphere available for the use of people as a whole (Ganetz 1995), then there is a clear contradiction between the common-sense understanding of the meaning of public space and the actual practice of the use, or desired use, of such places. Public space is the very space which adults want to keep children out of; they want to restrict children's access to, and activities within, it and consequently there is a distinctive shift in Western cultures towards the privatisation of public space (Aitken and Herman 1997; Davis 1990; Katz 1998; Valentine 1996a).

In the following section I introduce a particular micro-geography of the Rhondda Valleys in which economic change and stagnation, and strong aspects of community identity intersect in ways which mean that despite adult restrictions the streets are a significant part of teenage geographies. The remaining part of the chapter investigates the ways in which teenage girls construct 'maps of meaning' (Jackson 1989) through their interactions with three spaces: the public streets, the private space of the home, and a specific 'community' space. The chapter considers the adult definitions of public space and puts girls in spaces and places (public and on the streets) where existing literature on young people has not expected them to be.

Contexts of the research

The Rhondda Valleys

The Rhondda, with a population of about 78,000, is in fact two valleys, Rhondda Fach (little Rhondda) and Rhondda Fawr (large Rhondda) (see Map 5.1). In the late nineteenth century and for much of the twentieth century the Rhondda was a major contributor to the South Wales coal industry, and each small settlement along the Valleys was the site of a mine. In the 1980s and 1990s the region fell prey to global economic restructuring which has led to rapid deindustrialisation. The decline of coal-mining has forced the region into acute economic collapse; it is one of the poorest regions in the UK and qualifies for considerable financial support from the European Union Social Fund (see Table 5.1). Unemployment rates in the Rhondda for women aged 16–59 and men aged 16–65 are substantially higher than the national average at 10.23 per cent and 20.62 per cent respectively (the average for Britain is 6.98 per cent for women and 11.33 per cent for men). More women tend to find employment than men because of the growth in basic service industries and factory employment, much of which lacks security, is extremely poorly paid and invariably part-time.

The Rhondda has the second lowest level of student (aged 16–24) enrolment for girls in Wales and the lowest level for boys, while the unemployment rates for young people are amongst the highest in Britain. Hence the Rhondda is an area of considerable economic neglect and an example of an 'eroded

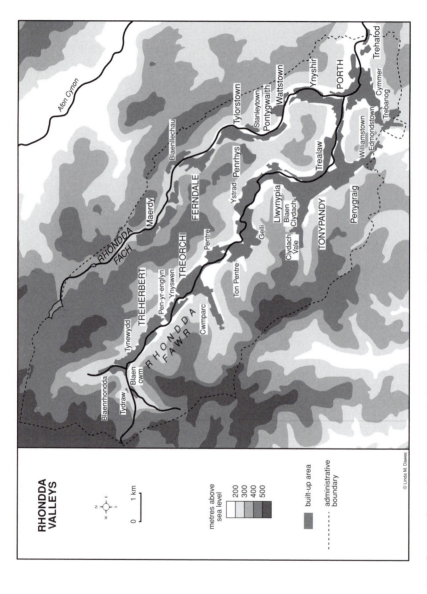

Map 5.1 Map showing the settlement areas which constitute the Rhondda Valley. Communities which are recognised as towns are shown in large capitals. The ways in which communities merge together can be clearly seen.

Table 5.1 Household resources and amenities for the Rhondda Valley, comparison of Welsh and British averages (all figures are shown as percentages)

	Rhondda	Wales	Britain
Lacking/sharing bathroom, shower and WC	4.35	1.74	1.25
No car	45.12	32.26	33.35
No central heating	23.91	18.54	18.88
No economically active adults	8.43	6.29	4.46
Percentage of residents in households with a member with a long-term illness	25.77	16.37	12.35

Source: Adapted from the 1991 Census data

ecology' (Katz 1998). The Rhondda lacks resources for public services including investments in education, training and public resources. When communities are poor, the children of such a community often suffer most of all.

Penygraig Community Project

Penygraig Community Project (PCP) is spatially located in two buildings in the centre of the small community[2] of Penygraig. Both buildings have been renovated and adapted by the project over the fifteen years it has existed. The largest building is the Soar Ffrwdamos, a former chapel. It has been converted into a multiple-use community space by the people residing in Penygraig and neighbouring communities and through funding from charities and local authorities.

PCP is an independent voluntary organisation and a registered charity. The project has a wide range of activities which are designed to meet community needs and are established through close relationships between Penygraig members and the wider community. Local people are involved in all aspects of the project, both as volunteers and in the management of the overall project. More than forty people at any one time are involved each week in running all the project's activities.

Although the project provides resources and service for all ages, it has a special focus on children and young people because of the level of disaffection experienced by the young, which is apparent in local statistics for drug use and crimes such as car theft. The PCP is firmly committed to challenging these social and cultural responses among children and young people, and to countering the stereotypical representations of the Rhondda youth. Their strategic aims are to help children and young people reach their potential, to support them and to assist them to develop knowledge and skills. The project is defined as educative, participative, empowering, in partnership, building relationships and expressive.[3] Activities for young people include play schemes (summer holiday activities for young children at which teenagers volunteer), a youth club, drama (Rhondda Youth Theatre), 'Stay Awakes' (children are sponsored to stay awake for the whole night, which they spend in a community building playing games and watching videos), 'Dragon Wheels' (a driving project geared towards preventing 'joy riding'), summer fêtes and summer camps.

PCP occupies a central position within a local geography and draws upon that geography to constitute itself and its activities. The project is funded at national and international levels (such as through the Lottery Charities Commission and the European Social Fund) and is also connected with children and youth projects in particular in other parts of Europe. It hosts volunteers working with the United Nations Youth Scheme who come from all over the world. Thus national and international groups play a role, through funding and exchange of expertise, within a localised and micro-geography of the Rhondda Valley. Hence PCP's own geographical activity spheres (Massey 1995) are tightly woven into the local community and expanded outwards internationally. There is a binary (we might even say mirror reflection) of neglect and resistance at play here; global restructuring causes economic collapse but local resistance taps into international networks in order to bring alternative patterns of training, support and care.

Working with teenage girls

The chapter is based on fieldwork conducted while I was working as a volunteer at PCP. With the help of one of the key youth co-ordinators I was able to conduct focus group discussions with a changing group of eight to ten teenage girls. We had several meetings through a five-month period, at which discussions were taped and transcribed. The fact that the timing of the discussions was structured by the girls themselves was empowering for them but frustrating for me sometimes, because I did not have as much time with them as I wanted. However, as part of the philosophy of the project it was important that the research be driven by the girls themselves in terms of when they wanted to let me ask them questions (Alderson 1995). What are presented here are the verbatim discussions with the girls. Interruptions, discontinued flow and contradictions are included, because the multiplicity of these voices and their ideas are important and central to an understanding of the girls' use of spaces and places (Madge et al. 1997; Opie 1992). The girls' names have been changed but I have used 'English' names because most of the girls had Anglicised names and none of them spoke Welsh as a first language.

Complexities of the spatial experiences of Rhondda teenage girls

Girls and 'public' space

There are very few commercial leisure spaces in the Rhondda (which in any case teenagers could not afford to use). There is only one cinema and no bowling allies, roller rinks, skating rinks or video and computer games centres, and the two sports centres in the region have few organised activities for girls. The streets and parks are therefore the only public spaces where Rhondda girls can spend their spare time. Like the boys featured in Paul Corrigan's (1979) study of young men in Sunderland during the late 1970s, the Rhondda girls are not driven out onto the streets purely because of the lack of alternatives available to them, but also make a positive choice to be there. In this way, the girls' use of the street disrupts Corrigan's notion that it is only a place for boys.

Although the street is a free space, which can at times be defined by children as their own, it is also a space where adults expect young people to show deference to adult definitions of appropriate behaviour (Valentine 1996a). The noise of teenagers, the shouting, laughing, screaming and constant movement of children 'hanging around', interrupts adult private domestic spaces, particularly in the Rhondda where the majority of the terraced houses face onto the street and the main door to the house opens straight off the street into the living room.[4] As a result, adults frequently threaten to call the police if the girls do not go home.

CAROL: If we're out on the streets then we get into trouble . . .

RACHEL: Not really trouble, it's just that everyone would be complaining and that, say we are causing trouble, they say we are making noise and all that.

LARA: They would blame us for doing something bad when we haven't done anything at all, they would want us to go away, so we're better off to come down here.

INTERVIEWER: What do you think about people's reactions when they ask you to move away, do you understand it or get fed up with it?

LOU: We get fed up because we think they are old moanies, we argue back sometimes, we swear at them, I won't do it now on tape, but we argue back and tell them we aren't doing anything wrong . . .

LARA: But they say 'Don't keep hanging around, we'll call the police' but it gets worse then because they usually end up calling the police . . .

CAROL: They don't always come though . . .

TERESA: They come then and they tell us we have to move or they tell us to go home.

SUSIE: They don't really care where we go, they don't expect us to be able to go anywhere.

CAROL: We just have to go to another corner don't we?

KEELY: There isn't anywhere else to go, there's only parks and that for us to go to and they do shut them early, the parks especially in the winter so you can't be anywhere . . .

Despite adults' best efforts, the girls persistently resist attempts to control their movements. They might give a particular corner or part of a park a wide girth for a few weeks until things 'calm down', but they are still out on the streets. In the summer when PCP is closed, they walk miles, enjoying each other's company and 'checking out what's going on'. Hence if they caused annoyance it was not usually on the same corner, near the same houses night after night, as it could be in the winter. They tend to keep within community spaces they know, but through school catchment geographies they know lots of teenagers in the surrounding communities. When they go out together in groups of up to ten they do not feel afraid and argue that they can resist adult attempts to limit their movements:

SUSIE: In the summer? Just go out hanging about.

LARA: Just walking around, walking up and down, looking, we walk about different roads.

CAROL: There's about ten of us.

KEELY: Most people then don't take any notice of us.

LOU: Lots of people are used to us round here.

SUSIE: Yeah that's how it is.

This assertion of 'that is how it is' implied that the girls felt they had every right to be out on the streets, meeting their friends and 'hanging around'. Thus through their acts of resistance the girls create the street more truthfully as a 'public' rather than an adult space.

Girls and private space

The growth in low-paid, low-grade work in the Rhondda means that women's working days have been extended, as they now undertake paid employment but still remain solely responsible for the daily running of the household. In all of the homes of the girls I interviewed, the gender division of labour was clearly demarked and the burden fell on their mothers and increasingly on them as teenage girls. This was one of their reasons for 'getting out of the house' so much and as often as they could. They were not ready to take on domestic responsibilities – although they would help their mothers out, they greatly resented having to wait on their brothers and fathers, as this excerpt shows:

CAROL: My brother doesn't do anything at all, and I don't! But my sister, she's 11 and she loves to do the cleaning. She's not right that girl! [Laughter]

KEELY: In my house my mother will say, 'Do the dishes then, Kee' and I say 'What about him?' and she will go, 'Oh well he's a boy' and I go, 'So?' Only sometimes it's like that, not all the time to be fair to my mum, some-times there is a change and he's got to chip in because he's older than me. If it gets too much though, I just leave, I don't stay and do the work.

CAROL: My brother doesn't live in our house but when he comes up at the weekend I have to do everything for him, "get me a drink, get me a biscuit', he's 22! I don't like to have to do all of that, I think if he wants he can get it himself. He assumes he can come back and I will do it all for him, because I'm younger. I tell him to do it himself but he won't have it then, he just sits there. In the week he lives with my father and I think he must do things for himself because he works. When I go down to my father's each week then I usually do the dishes and things, sometimes I do the ironing for him . . .

In her study of young women's transition from school into the job market, Christine Griffin showed that girls were strongly socialised at that time into helping their mothers in the domestic sphere. In her study she found that only 8 per cent of fathers and none of the girls' brothers did any domestic work (Griffin 1985: 37). She found that white middle-class girls were the least likely to be expected to help with housework. The girls in my study all came from working-class or work-poor households and yet somehow managed to resist the amount of housework Griffin's teenagers had to do. The way to avoid the work was sometimes through an appeal for equality to their mothers, sometimes through direct refusal in the case of demands from their brothers, but mostly it was to get themselves out of the house.

David Sibley's analysis of families and domestic routines examines the complex ways in which children experience social space and the ways in which bound-aries (invariably created by adults) are an integral part of children's lives in both the home and the 'near-home environment' (Sibley 1995b: 125). Sibley (1995b: 129) shows that the home 'is one place where children are subject to controls by parents over the use of space and time' and that the actual 'availability of space in the home may limit opportunities for children to secure privacy'. Most of the girls in the study either shared a bedroom with their sister or had a very small bedroom to themselves. The houses in the Rhondda are quite small, based on the two-up two-down principle, although most have had extensions to provide inside bathrooms and often a third bedroom. As the houses are small, they leave little room for child-only spaces. Therefore, if faced with demands for domestic work or embroiled in some kind of adult–teenager conflict, the girls can't go to their own rooms and have time alone and away from adult interference. Instead they will leave the space of the home and find their friends and spend the rest of the time on the street. Most of the girls stayed out until 10.30 p.m. on certain weekday nights and would go back home, or to a friend's to watch the soap opera *Brookside*, which was the first non-Welsh language programme of the evening on S4C (the Welsh equivalent to Channel 4, which is a British commercial channel with a special remit to cater for minority audiences). Rather than a 'retreat from the streets' (Valentine 1996b: 586), the girls 'escape from the home' and use the public streets as an escape route from the domestic responsibilities the inner-city working-class girls in Griffin's study had to accept.

Throughout the weeks of meetings and discussions none of the girls talked about their bedrooms and being in each other's rooms as part of their experience with friends (Ganetz 1995). In contrast to the working-class girls of McRobbie's study, for the girls in the Rhondda, family and domestic life were not of central importance in their lives. They had ambitions to travel, one talked about being a holiday courier in Spain, and none of them talked of marriage or having children (McRobbie 1991: 53). The girls at this age, although their economic and education options are curtailed because of deindustrialisation and decline, still have hopes of work and possibilities of spending time away from the Rhondda and are not in the least fixated with romantic notions of marriage, motherhood and domesticity.

Girls' experiences of the Penygraig space

In lots of ways the Penygraig space occupies an ambiguous position on the public/private binary. The project is for the community and it is a site of work and production for many adults. It is also open to public scrutiny and accountability in relation to funding and legislation relating to working with young people. Hence it can be defined as a public space. Alternatively, there are periods of time when the designated use of the space and the limits of entry are quite distinct – usually based on age, sometimes on gender. The relationships which are established over periods of time between young people themselves, and between the children and the youth workers, are often very personalised. For some of the children the clubs they attend are like a second home, they can get something to eat and drink, they can watch television and play games. Consequently the sociality of the space can make it feel like a private space. It is a good example of the ways in which people make places and spaces and because of the social relations within which such spaces and places are located they can appear at once public and private (Massey 1995). This fluidity of meaning of the PCP means that the girls could be both positive and critical of the space. There was an ambivalence about the space which reflected their age, not still children and yet not yet adults, not yet sure of where they can be and what they can do. They talked about what it meant to them as it functioned now, but also in terms of what they wanted it to be able to provide.

INTERVIEWER: So is Penygraig a good place to come or is it because there is nothing else?

TERESA: Nothing else to do.

SUSIE: I don't mind it here though, I like coming here.

CAROL: I'd rather if they run trips, you know take us places, skating and stuff like that.

KEELY: It doesn't always have to be something really special – just nights out and things like that really.

LOU: Because when you come down here it's like doing the same thing every night.

LARA: They could take us swimming or bowling, stuff like that.

KEELY: Because if it's just club and thing like this here – they you know, people just won't go because they get bored of it.

[edit]

SUSIE: When we had the air bed, I mean the bouncy castle, then it used to be full, no-one wants to come now that it's gone.

The loss of the inflatable castle (the boys were too rough and punctured it) and the ways in which it created a different space to be within the youth club space appeared repeatedly throughout the interviews.

SUSIE: I love the castle I do, everyone likes coming when we have it.

LARA: I'd stay on there all night if I could, it's just so good to jump around and that.

[edit]

RACHEL: The little kids love it but we just want to sit down on it, at the front or in a corner and be able to talk there, it's just different than sitting in the normal places, on the chairs. In the castle it's different.

For the girls the bouncy castle had made the space of Penygraig different because none of the other clubs had anything like that. It allowed them to play and also to sit and talk.[5] Ganetz argues that:

> The relationship orientation which female friendship entails may be expressed in several ways, *inter alia* through the establishment of private zones in public spaces, which is so typical for young women.
>
> (Ganetz 1995: 85)

The castle provided just such a zone for the girls, creating an island of privacy where they could sit and talk within a space dominated by boisterous boys.

Likewise, the girls also have the opportunity to shape the space of the project on disco/club nights. The PCP regularly arranged dance nights, allowing the older children to be involved in the organisation of the evening, for example choosing the music, using the equipment and so on. The space of the dance floor was gendered especially around particular songs, but unlike the non-dancing boys of McRobbie's study (1991) the Rhondda boys danced either with the girls or independently. Though the girls who are 16 find the Penygraig nights too young for them (they prefer to try and pass for 18 and get into a nightclub called 'The Bank'), for the younger girls the dance night is something they want to be at, although they in turn resent the presence of pre-teen children younger than themselves.

INTERVIEWER: So who came on Friday?

SUSIE: I did, but there were little kids running around, sometimes they were about 7 or 8, running around small kids, some of them are young, my sister's friend, she's 9 and she was here!

LARA: There were a few older people too which was good.

SUSIE: Yeah, people from Ystrad, they are from my school, they are about our age.

TERESA: I like all the music they play I do, I like the disco and I like rave.

Unlike the youth club culture of youth workers described by McRobbie (1991) and Nava (1992), Penygraig presented much more flexibility and autonomy for the teenage participants. It allowed them to create their own spaces and to choose how to behave within them. There are of course established boundaries within the youth club, but there were only three rules which the children all knew and on the whole followed: 'Respect each other, Respect the place, Respect yourself'. Such principles were drawn from discussions between adult workers/volunteers and the children themselves as a workable and fair baseline of behaviour everyone could expect from each other. The concept of empowerment is an important one in Penygraig's philosophy (personal communication, Pauline Richards). I witnessed no fights, severe verbal aggression or vandalism in the months I worked at the project. If tempers began to rise, individuals were either taken to one side and talked to by a worker or volunteer or the whole group would be stopped until quiet and would be asked to think about what they were doing and why it might be causing a problem. Penygraig is a project of and for the community, organised and maintained by the community in which it is located. Firmly rooted within its local social and cultural geography, it sees teenagers as an integral part of the project and as potential future members of the project.

For this group of girls, the youth club space of Penygraig is many things. It is at once a good place to be, with things happening which they enjoy, but it is also somewhere to go rather than be on the streets, outside in the rain. They participate in the activities set up for their age group and they show their commitment to the project by attending most weeks and in some cases volunteering for the summer play schemes. However, they have clear ideas of what they feel they are missing out on. They would like to go to commercial leisure spaces, the types of places many children and young people from more privileged backgrounds go to on a regular basis and perceive to be part of their culture.

However, with or without Penygraig, teenage girl friendships create strong senses of belonging and identity. The girls I talked to valued the project because of the friendships it had helped them build. Through these friendships they can participate in the project but they can also maintain an active geography in the public spaces of the streets and the parks.

TERESA: That's the good thing about coming here, to the Project you meet your friends. Like I met Keely through drama and then I met Carol through Keely, and the same with Lou.

KEELY: If I hadn't come to drama I wouldn't know any of these, none of them, no body.

SUSIE: Yea, like they are older than me and I wouldn't have got to know them with coming here – now they are my best friends.

RACHEL: I have other friends I see but this crowd, when we come here on Thursday's we make it good for all of us.

CAROL: It's a good laugh it is, yea, we have a really good laugh some of the time, we can just giggle and be together here, there's no trouble, you can just be yourself.

Conclusion

The Rhondda Valley is an example of an 'eroded ecology' (Katz 1998), which means that the lives of children are impoverished and lacking in opportunities, diversity and activities. Global and economic shifts and transformations have rendered the region socially and economically poor. However, through community projects like Penygraig, residents of the regions are creating positive and welcoming spaces for children. It might not be all these teenage girls would like it to be, but it is a place for them, somewhere they feel central and important. It provides an alternative to some of the stifling domestic settings and the contested spaces of the streets. Without a place like Penygraig, these girls would spend most of their everynight lives on the street. Being on the street brings them into much more regular contact with activities which the Penygraig space allows them choices to keep away from: negative youth activities such as car theft, joy riding, drug dealing and violence. The girls know that such things happen, and they come into contact with them when they are on the street, but their involvement with Penygraig means that they do have an alternative and that within the youth club space they create their own social worlds which revolve around friendship, being together and 'chatting'. They then use these connections to make good use of the streets as a social space rather than a place in which only the negative elements of deprived youth experiences happen. Their best form of leisure is being with each other and the fun they share through common geographies, lifestyles and gender.

My work with Rhondda teenage girls is important because it challenges and expands much of the existing material which focuses on this age group of children (Corrigan 1979; Fornäs and Bolin 1995; Furlong and Cartmel 1997; Griffin 1985; McRobbie 1991; McRobbie and Garber 1976; McRobbie and Nava 1984; Nava 1992). The girls were not obsessed with idealised romantic notions nor were they already shackled into training for domestic 'labours of love'. They actively worked at being out and about, at 'hanging around' and participating in a dynamic and public teenage girl culture based on the streets and in the project youth clubs. Each time we began to discuss their experiences and thoughts about key institutions in which they participated – school, college, home – their conversations were always quick to move onto stories of the street, last week at the youth club, meeting up at the weekend. The girls' social and leisure geographies were active, significant, and clearly positive and central features of their lives. The ways in which the girls could maintain their active geographical presence was through their friendships and the close networks they built up between themselves. In some cases these friendships were reinforced through contact at school, but for most of the girls I worked with, their times and places to meet and enjoy each other's company was always on the street or at Penygraig. Going about in groups of three or more gave the girls a security of company and support which they valued and recognised as a key factor in allowing them to escape duties at home, to be able to go to different places within walking distance and to feel comfortable and 'at home' in the Penygraig Project spaces. The ways in which friendships intersect with, and create, children's geographical experiences are a crucial area of geographical study. For the Rhondda girls, being a 'Valley Girl', 'having a laugh and that', was their favourite pastime and such friendships allowed them significant freedom and mobility in an otherwise deprived and impoverished environment. I repeat Carol's statement about why her girl friends were important:

> It's a good laugh it is, yea, we have a really good laugh some of the time, we can just giggle and be together here, there's no trouble, you can just be yourself.

Acknowledgements

My very deep thanks go to all the people at the Penygraig Community Project for their support, encouragement and interest in the research I did. I would especially like to thank Pauline Richards who has been, and continues to be, an inspiration in her commitment to young people and her dedication to her community, not to mention being one of the funniest people I know. I would like to thank all the girls who participated in the research and who shared parts of their lives with me. Thanks to Sue Loveridge who introduced me to the Penygraig Community Project in the first place. I would also like to thank Sarah and Gill as editors for their patience and very useful advice on an earlier draft of this chapter.

Notes

1 I deliberately use the term 'everynight' inspired by Malbon (1998).

2 I use the term 'community' here for specific reasons. I do not wish to imply any form of homogeneity amongst the people who reside in these places, but rather use the term for some kind of geographical accuracy. The named settlements which run the length of the two Rhondda Valleys collectively form what is called the Rhondda Valleys, the Rhondda or the Valleys. Each named place marks the place of a coal-mine and the houses of the miners and their families which clustered around it. To indicate that the named places are not and never have been villages, and few are large enough to constitute towns, I use the term 'community'.

3 These definitions are taken from the Penygraig *Working with Young People* pamphlet published in 1997.

4 In this discussion I am not presuming the 'home' to be always a place of safety and protection. The 'home' of course can be a very dangerous and abusive place for children (Rose 1995; Women and geography Study Group 1997).

5 With the participation of the community, including children themselves raising money through their 'stay awakes', Penygraig was able to replace the castle at the end of 1996.

References

Aitken, S. and Herman, T. (1997) 'Gender, power and crib geography: transitional spaces and potential places', *Gender, Place and Culture* 4, 1: 63–88.

Alderson, P. (1995) *Listening to Children: Children, Ethics and Social Research*, Ilford: Barnardo's.

Boëthius, U. (1995) 'Youth, the media and moral panics', in J. Fornäs and G. Bolin (eds) *Youth Culture in Late Modernity*, London: Sage.

Corrigan, P. (1979) *Schooling the Smash Street Kids*, Basingstoke: Macmillan.

Davis, M. (1990) *City of Quartz*, London: Verso.

Fornäs, J. and Bolin, G. (eds) (1995) *Youth Culture in Late Modernity*, London: Sage.

Furlong, A. and Cartmel, F. (1997) *Young People and Social Change: Individualisation and Risk in Late Modernity*, Buckingham: Open University Press.

Ganetz, H. (1995) 'The shop, the home and femininity as a masquerade', in J. Fornäs and G. Bolin (eds) *Youth Culture in Late Modernity*, London: Sage.

Griffin, C. (1985) *Typical Girls? Young Women From School to Job Market*, London: Routledge and Kegan Paul.

Holstein-Beck, S. (1995) 'Consistency and change in the lifeworld of young women', in J. Fornäs and G. Bolin (eds) *Youth Culture in Late Modernity*, London: Sage.

Jackson, P. (1989) *Maps of Meaning*, London: Routledge.

James, A. (1986) 'Learning to belong: the boundaries of adolescence', in A.P. Cohen (ed.) *Symbolising Boundaries: Identity and Diversity in British Cultures*, Manchester: Manchester University Press.

Katz, C. (1998) 'Disintegrating developments: global economic restructuring and the eroding of ecologies of youth', in T. Skelton and G. Valentine (eds) *Cool Places*, London: Routledge.

Laurie, N., Smith, F., Bowlby, S., Foord, J., Monk, S., Radcliffe, S., Rowlands, J., Townsend, J., Young, L. and Gregson, N. (1997) 'In and out of bounds and resisting boundaries: feminist geographies of space and place', in Women and Geography Study Group, *Feminist Geographies*, London: Longman.

Lucas, T. (1998) 'Youth gangs and moral panics in Santa Cruz, California', in T. Skelton and G. Valentine (eds) *Cool Places*, London: Routledge.

Madge, C., Raghuram, P., Skelton, T., Willis, K. and Williams, J. (1997) 'Methods and methodologies in feminist geographies: politics, practice and power', in Women and Geography Study Group, *Feminist Geographies*, London: Longman.

Malbon, B. (1998) 'Clubbing: consumption, identity and the spatial practices of every-night life', in T. Skelton and G. Valentine (eds) *Cool Places*, London: Routledge.

Massey, D. (1995) 'The conceptualization of place' in D. Massey and P. Jess (eds) *A Place in the World?*, Oxford: Oxford and Open University Presses.

Matthews, M.H., Limb, M. and Taylor, M. (1999) 'Defining an agenda for the geography of children', *Progress in Human Geography* 23, 1: 59–88.

McNamee, S. (1998) 'Youth, gender and video games: power and control in the home', in T. Skelton and G. Valentine (eds) *Cool Places*, London: Routledge.

McRobbie, A. (1982) 'Jackie: an ideology of adolescent femininity', in B. Waites, T. Bennett and G. Martin (eds) *Popular Culture: Past and Present*, London: Croom Helm in association with the Open University Press.

—— (1991) *Feminism and Youth Culture*, Basingstoke: Macmillan.

McRobbie, A. and Garber, J. (1976) 'Girls and subcultures', in S. Hall and T. Jefferson (eds) *Resistance Through Rituals: Youth Subcultures in Post-war Britain*, London: Hutchinson.

McRobbie, A. and Nava, M. (eds) (1984) *Gender and Generation*, London: Methuen.

Nava, M. (1992) *Changing Cultures: Feminism, Youth and Consumerism*, London: Sage.

Opie, A. (1992) 'Qualitative research, appropriation of the "other" and empowerment', *Feminist Review* 40, 1: 52–69.

Penygraig Community Project (1997) *Working with Young People*, Penygraig: PCP.

Rose, G. (1995) 'Place and identity: a sense of place', in D. Massey and P. Jess (eds) *A Place in the World?*, Oxford: Open University and Oxford University Presses.

Sibley, D. (1995a) *Geographies of Exclusion*, London: Routledge.

—— (1995b) 'Families and domestic routines', in S. Pile and N. Thrift (eds) *Mapping the Subject*, London: Routledge.

Skelton, T. and Valentine, G. (eds) *Cool Places: Geographies of Youth Cultures*, London: Routledge.

Valentine, G. (1996a) 'Children should be seen and not heard: the production and transgression of adults' public space', *Urban Geography* 17, 3: 205–20.

—— (1996b) 'Angels and devils: moral landscapes of childhood', *Environment and Planning D: Society and Space* 14: 581–599.

Valentine, G., Skelton, T. and Chambers, D. (1998) 'Cool places: an introduction to youth and youth cultures', in T. Skelton and G. Valentine (eds) *Cool Places*, London: Routledge.

Women and Geography Study Group (1997) *Feminist Geographies: Explorations in Diversity and Difference*, London: Longman.

6

TIME FOR A PARTY!

Making sense of the commercialisation of leisure space for children

John H. McKendrick, Michael G. Bradford and Anna V. Fielder

Commercialisation of children's playspace

There are spaces that we readily associate with children, the most prominent of which are the school, the home and the (near) neighbourhood (van Vliet 1983).[1] An integral part of each of these domains are designated sites for leisure. The school has its playground, yard or sports field; the home has its private bedrooms, garden and increasingly a playroom; and the neighbourhood has its community centre, public open spaces and playgrounds. Traditionally, these sites have been nodal points of a contiguous expanse of space within which children play. Independent access to this playspace is initially centred on the home and expands outward from it at differential rates according to age, co-presence of peers and gender (Hart 1979; Moore 1986; and Matthews 1987, respectively).

In recent years many children's independent access to neighbourhood (play)space has been curtailed in the North as adults react to fears for their children, and to dangerous children (Valentine 1996). Explanation for more interventionist parenting has been attributed to stranger-danger (Blakely 1993; Valentine and McKendrick 1997), increased traffic flows that render streets dangerous for play (Davis and Jones 1996), fear of paedophilia, child abduction and murder (McNeish and Roberts 1995), unruly children (McGallagly *et al.* 1998), and the 'appropriation' of traditional playspace by planners (Roberts, Smith and Bryce 1995). Exacerbating these threats to playspace, are the resource constraints which now threaten public parks and their playgrounds (Webb 1997). However, the recent history of children at play in the North is not merely a trend towards curtailment and impoverishment.

In the advanced industrialised countries of the North the disposable income of many families has increased to unprecedented levels, contributing towards the diversification and commodification of play in the homespace (McNeal 1992; see also Chapter 9). Yet with respect to the public geography of children at play, these new developments in many ways *reinforce* the trends towards restricting children's presence outdoors. For example, commodified home-play offers a pull-factor to exacerbate the push-effect that seeks to 'keep young people off the streets' (Jeffs and Smith 1996). And although flagship community youth-space developments (South Ayrshire Council 1999) encourage young people to use their local area, access remains subject to stringent conditions of time and space.

In this respect it can therefore be argued that the most significant development to challenge the prevailing order is the development of commercial playgrounds for young children (McKendrick, Fielder and Bradford 1999). The 1990s expansion in the provision of 'add-on' indoor and outdoor playgrounds in motorway service stations, airports, hospitals, ferries, banks and building societies, family pubs/restaurants,[2] retail outlets and supermarkets, and the provision of 'stand-alone' indoor soft-play centres, have asserted children's right to playspace in parts of the built environment which were hitherto perceived as almost exclusively adult domains.

'Commercial' playgrounds is a descriptor that encompasses a range of play environments. Even their common points of reference – the commercial dimension and the playground environment – differ significantly across sites and genres (McKendrick *et al.* 1999). Even so, it is instructive to perceive these as commercial playgrounds in that they generate profits by offering opportunities for play in a designated and safety approved site. Most commercial playgrounds are situated indoors. Adult viewing areas are located alongside the play zones. Many of these play zones, particularly those in non-leisure domains, are aimed at a 'younger' age group. Manipulable soft-play shapes, small ball pools, domestic replica equipment and vehicular replica equipment are typical features in small, enclosed spaces. The most extensive playgrounds tend to be multi-level, to replicate the features of small indoor soft-play centres on a larger scale, and to supplement these with rope swings, net ladders, and a variety of slides. Family pubs/restaurants often provide an outdoor playground, consisting primarily of a multi-functional climbing frame on a 'safe surface' to complement its indoor soft-play centre. Equipment tends to be fairly standardised, reflecting its modular construction and equipment supply networks. However, the equipment is packaged and themed to create a brand identity, for example *Alphabet Zoo, Jungle Bungle* and *Wacky Warehouse*. Children are 'free' to interact with one another and the equipment: parents are generally held responsible for their children's behaviour and safety, but are permitted limited access to the actual play environments.

The commercialisation of playgrounds raises challenging questions. To what extent are these new opportunities for play an integral part of children's lives?

To what extent do they celebrate the child and meet her or his needs? In this chapter, these issues are explored through the medium of children's birthday parties. As these parties are in themselves celebrations of the child, and as commercial playgrounds are increasingly the sites in which children's birthday parties take place, this is a fruitful point of engagement. We begin by briefly introducing the research project upon which the chapter is based. We then move on to discuss patterns of birthday party participation, and the changing social and spatial functions of the party. In the conclusion, 'The party's over', we consider the wider significance of the commercial provision of leisure space for young children.

The business of children's play

The research upon which this chapter is based derives from The Business of Children's Play research project which ran from 1996 to 1998 (McKendrick, Fielder and Bradford 1998). The study reviewed various aspects of the usage, provision and consumption of commercial playspace, focusing on the Greater Manchester area. The research was designed as a multi-method, multi-stage project (McKendrick 1999), which adopted a strongly child-centred focus. Six commercial providers of children's playgrounds were profiled, and research was conducted at ten field sites including four schools. Nearly 900 families (both users and non-users) were surveyed with the prime intention of establishing children's usage of the centres. Thirty of these families were interviewed (adults and children separately) regarding their perceptions and feelings about the centres, and to establish the implications of the play centres upon other signif-icant domains in children's lives. Seven of these families participated in day-visits followed by a video-based interview. This part of the research provided detailed case studies of children at play in one commercial playground. However, key themes were explored across these methods, one example of which was that of children's birthday parties.

Party time!

Opportunity knocks

The birthday party is a major calendar event in the year of many children, and it is one that transforms domestic or commercial playspace for a fixed and limited time (Plate 6.1). In terms of the Christian calendar, for example, it is the most personal of all annual celebrations, with others being orientated towards the whole cohort of which the child is part (e.g. Easter and the 'modern tradition' of gifting chocolate eggs). In contrast, the birthday party is in many cultures the

Plate 6.1 Children's birthday party in the home. © John McKendrick

event that is most closely associated with a celebration of the individual child, that celebration being bound up with society's preoccupation with age in the context of child development (Bentley 1998). The birthday party is also pertinent to broader debates on commodification and commercialisation in children's lives (Kline 1993; McNeal 1992). On the one hand, it is not necessarily the most 'rewarding' event in terms of the value of consumer goods bestowed on the child. For example, British children who celebrate Christmas tend to receive more, and more expensive, presents at that particular celebration. Furthermore, the child in preference to a present often chooses the party, or a smaller present is received than would otherwise have been the case. In this respect, the children's birthday party can be understood as a bulwark against commodification of play which upholds, in preference, a social gathering of friends and family in which 'party' rituals are partaken (birthday cake, birthday songs and party games). Yet, on the other hand, it is increasingly an event which is conducted outside the family home and in the domain of commercial playgrounds.

The research established that one of the key reasons that children visit indoor soft-play centres and family pubs/restaurants is to attend a children's birthday party: seven-tenths of children who had visited an indoor soft-play centre, and three-fifths of children who had visited a family pub/restaurant had done so, on at least one occasion, to attend a birthday party (Table 6.1). Indeed for indoor soft-play centres almost as many visited to attend a birthday party as had visited in the company of family (70 per cent and 74 per cent respectively). Unlike the theme park (with its regional catchment area) and the shop with a play zone (in which 'play' is a subsidiary and supporting function), the family pub/restaurant and indoor soft-play centre serve local and metropolitan catchment areas and have taken steps to capture their share of this market. It is a sizeable market:

> MR DOUGLAS: Yeah, [the number of parties catered for in a day] varies . . . Southwood,[3] for example, can have fourteen on a weekend in a day. So it's like twenty-eight over the weekend . . . And then we can have two a night. Some of the centres we fit in about eight to ten on a weekend. And then again, two to three in the evening.
>
> (Area Manager, chain of indoor soft-play centres)

For the service provider, the attraction of birthday parties not only rests in the number of customers who are brought directly to the facility: birthday parties also introduce potential new customers (i.e. children who have been invited to the birthday party), which holds potential for future repeat visits of these children with their families and for their own birthday parties.

Decision-making

Parents gleaned information about commercial playgrounds from a wide variety of sources, of which friends/family (37 per cent), when driving/walking past

Table 6.1 Ever visited a commercial playground for a children's birthday party: genre by school

Commercial playground	Percentage of survey children who have visited for a birthday party				
	School				All-School
	Lancastarian	Ewings	Firs	Woodheys	
Indoor soft-play centre	33	30	60	87	70
Family pub-restaurant	13	32	49	76	59
Children's theme park	14	4	6	7	6
Shop with play zone	0	8	4	2	3

Source: Authors survey via schools

Cases: Variable, ranging respectively for each school 11–15, 22–8, 100–12 and 136–48

Notes: Data were drawn from a multiple response question in which guardians/parents were asked to indicate *all* the groups/visit parties of which the child was part on a visit to each of the four genres of commercial playground. 'Children's party' was one of six options presented to the respondents. The size of the survey population for Lancastrian School, in particular, necessitates careful interpretation of these data.

(25 per cent) and advertisements (18 per cent) were the most important. One-eighth of parents acquired their knowledge about commercial playgrounds through their son or daughter being invited to a birthday party. As Table 6.2 demonstrates, the birthday party was relatively more important as a source of information for playgrounds with more localised catchment areas, that is for indoor soft-play centres as opposed to family pubs/restaurants.

However, while parents' awareness of commercial playgrounds may, at least for some and at least in part, be attributed to their child attending a birthday party, the decision to host a birthday party in a commercial playground is mainly the preserve of the parent:

> LORRAINE: [The decision] was mine really. I just thought his friends are still very young, and just to try and entertain them for two hours would just be virtually impossible. So, I had to sort of talk him into it to be honest . . . he [William] wanted a bouncy castle in the garden.

Table 6.2 The children's party as a means of raising awareness of commercial play centres among parents

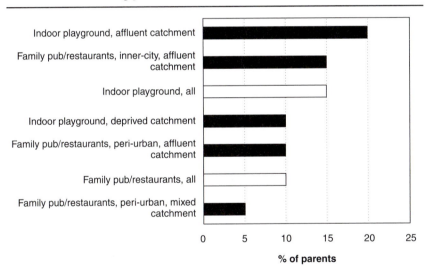

Source: Author's survey in play centres

Cases: 541 gross, of which in descending order of bar 92, 75, 165, 73, 70, 227, and 82, respectively

Note: Data were drawn from a multiple response question in which guardians/parents accompanying children to commercial play centres were asked to indicate *all* their sources of information about the centre in which they were surveyed. 'Children's party' was one of eight options presented to the respondents.

He'd had one last year and it rained last year and it was a disaster, so we talked him out of that.

(lone mother, two children aged 3 and 6)

It would be hard to find fault with parents for making such decisions on the grounds of seasonal functionalism (to avoid outdoor parties in the unsuitable climes of a winter in the northern hemisphere) or the life course of a child (to ensure that the party experience is age-appropriate). However, as this parent suggests, such reasoning is bundled together alongside the need to minimise inconvenience to parents and because parents lack the ability to engage their child and her or his guests for the duration of a party. With respect to decision-making, there are signs that the birthday party experience is not as child-orientated as one would have been expected.

Life course perspectives

The patronage of commercial playgrounds for birthday parties also marks a drift away from parties in the homespace. This is demonstrated by Hazel, who has had all recent parties in a commercial playground, and Colin who, according to his mother, has little experience of attending parties in domestic settings:

HAZEL: . . . all the parties that I've had . . . my sixth and my fifth I've had at Kidz Kingdom, and my fourth one I've had at the PlayZone.

(girl, aged 8, middle-class family)

MARION: Just thinking about *traditional* parties . . . This is Colin's fourth year of going to school, and I think he's probably gone to *two* parties at people's houses. That's all, out of what, twenty or thirty?

(mother of four children aged between 3 to 8, two-parent, middle-class family)

This transition from the domestic party setting to the commercial setting tends to occur at the age when the child enters pre-school education, that is as the child's social world becomes orientated towards the peer group (compare Tables 6.1 and 6.3). The transition is not merely a change of venue: family, particularly extended family but also to a lesser extent the parents, play a less prominent role in the party and 'traditional' party games and rituals are less prevalent. The environment of the commercial playground – noisy, fast and busy – is less comforting to older generations within the extended family; parents have a much more passive role in the proceedings, with less control over the activities and eating arrangements; much less time is available for 'traditional' party games with more time being spent at play; and more space is available for use by party participants, albeit the environment must be shared with other users. Even so, the basis for the party remains constant: the celebration of the individual child, the participation of close family and friends, and the sharing of traditional party practices.

The transition in the early years from a domestic to a commercial site as the domain for the birthday party, is not the only life course change that is evident for the birthday party. Just as young children grow into birthday parties, so older children imprint their own specific requirements upon them, for example, some resist playing the organised party games, or request party food at home rather than at the centre itself. And finally, as Donna explains, children eventually move on from the party in the commercial playground in the latter years of middle childhood:

> DONNA: Um, I think it's a bit naff now, when you're ten, you can't have a birthday party at Adventure World . . .
>
> (girl, aged 10, working-class family)

Thus, the drift of birthday parties from the homespace to the commercial playground is associated with a particular stage in the life course. Commercial playgrounds cater well for groups of young children; as children approach middle childhood, they tend to view the party activities surrounding play and, eventually, the whole play-party package less favourably. At this point, children tend to return to the homespace for 'sleepover' parties, or move on to other commercial facilities, where leisure or sport are the added attraction of the birthday party.

Something for everyone?

From a commercial standpoint, older children 'growing out' of birthday parties is not necessarily problematic. Given stable numbers (a market) of children between the ages of 2 and 10 to 12, as an older cohort of children grow indifferent to the play centre, a younger cohort takes its place. Problems would only arise if birth rates dropped below replacement levels, or if the market contracts as children grow indifferent to play centres at an earlier age: at this point, there is no evidence to support such trends. However, this is not to suggest that commercial playgrounds offer something for all children in early and middle childhood.

Returning to patterns of participation in general, and attendance at birthday parties in particular, we found that there is clear evidence of geo-social differences. As Table 6.3 shows, children from less affluent areas are nine times more likely not to have attended an indoor soft-play centre (9 per cent, as against 1 per cent from affluent areas), and almost twice as likely not have visited a shop with a play zone (15 per cent as against 9 per cent from affluent areas).

Though, on the whole, geo-social background is not associated with the likelihood of participation in family pubs/restaurants (about 3 per cent were non-attendees), marked differences were evident with respect to the likelihood of visiting a family pub/restaurant to attend a children's birthday party (Table 6.1): though only one-half of children from less affluent areas had attended for this reason (49 per cent), the equivalent proportion of children from more affluent areas was three-quarters (76 per cent). It is less clear whether these geo-social patterns are a direct function of the costs incurred:

> LORRAINE: my son went to [an indoor soft-play centre] for a birthday party. It's just too expensive for me, for the amount of time that they're there. It just gets far too expensive. Three or four pounds to get in, and I just can't afford that. But, I mean, . . . I could've stayed but you'd have to buy your own drinks and everything and I just couldn't afford to.
>
> (lone mother, of two children aged 3–6)

> JULIA: Yeah, if you have a party you have to do the play bit and then they do the food as well. There's a choice off the menu, you know, sausage, chips, burgers and that sort of stuff. And then you can bring your own cake or they'll do the cake. I mean it's really good, and it's brilliant value. You wouldn't do that at home for that.
>
> (mother of children aged 2 and 4, two-parent family)

Though for some parents from low-income families, the costs incurred are such that they are themselves unable to participate while their child attends a birthday party, let alone arrange for their child to host their party at such a

Table 6.3 Ever visited a commercial playground: genre by school

Commercial playground	Percentage of survey children never having visited School				All-School
	Lancastarian	Ewings	Firs	Woodheys	
Indoor soft-play centre	29	13	9	1	7
Family pub-restaurant	12	29	3	12	6
Children's theme park	18	50	8	9	9
Shop with play zone	35	16	15	9	13

Source: Authors survey via schools

Cases: Variable, ranging respectively for each school from 17, 31, 117 and 147–9.

Notes: Data were drawn from four different questions from author's survey. The size of the survey population for Lancastrian school, in particular, necessitates careful interpretation of these data.

venue, others find these to be a less costly party alternative. However, there is evidence to suggest that patterns of participation are shaped, if not by direct cost, then by other by-products of geo-social exclusion, for example mobility and transportation.

> CAROLINE: . . . the problem was that I *know* that some of the children in his class live some distance away, plus their parents haven't got transport, so the last thing I wanted to do was to invite the children, the children to know that they'd been invited, and then them have to say 'I can't come'. So I was a bit selective . . . I knew which parents would make the effort to bring their children.
>
> (mother of a 5-year-old with learning difficulties
who attends a 'Special School')

Turning more directly to questions of social groups and inclusion/exclusion also presents conflicting evidence. On one hand, there is clear evidence to suggest that children with disabilities are less likely to participate and that, among users, they are much less likely to have visited with the intention of attending a birthday party (Tables 6.2 and 6.1 respectively). For example, while one-third of users from 'special needs' schools had attended a birthday party at an indoor soft-play centre, the equivalent proportions for mainstream schools were three-fifths (less affluent area) and nine-tenths (affluent area). Then again, for other children, commercial playgrounds provided a space which facilitated social inclusion of some minority children:

> KUOMI: . . . it is difficult for us to invite other children to my home because we haven't got the faculty in English [*laughs*], so it's easier to invite them to Kidz Kingdom . . . then leave them.
>
> (father of two children aged 5–8,
non-UK ethnic origin)

Parents expressed more agreement on the question of the quality of the party experience at commercial playgrounds. Adults' opinions were generally *less* favourable than children's, with specific objections being raised about the quality of the food, the professionalism of the event, and the standardisation of the whole party experience. These are concerns with which the service providers are aware. Significantly, in formulating solutions to these acknowledged problems, there is a sense in which the problem is not inherent to the party package, and is not one that is acknowledged by children. Rather, as is suggested below, it is the parents who express concern, in particular with the presentation of the event:

MR DOUGLAS: It was, it was terrible. Yeah . . . Yeah, we changed the whole food around. Erm, now the presentation of the food is like a hundred per cent better. I mean it's not so much for the children because the children couldn't care less. It *is* for the parents. And now when a party table is laid up at a Kidz Kingdom now, you do go 'Ooh, wow, that looks really nice.' Erm, I mean we had like little white plastic I mean like vending cups. Now we've got like multi-coloured solid plastic cups, like different coloured spoons, plates, just the whole presentation is a lot better.

(Area Manager, chain of indoor soft-play centres)

Reconfiguring place

The changes being instigated by service providers are of recent origin and could not account for the already established popularity of their playgrounds as a venue for birthday parties. This raises the interesting issue of why parties have moved from the homespace, if parents are, on the whole, less than complementary about the party package with which they are presented. Part of the explanation for this paradox has already been alluded to: parties are as much for (the benefit of) parents, as they are for (a celebration of) the child:

MARION: Karen's last birthday was, the most recent one was at Kidz Kingdom, and it's so easy, it's like cheating . . . it is, really, the only thing that you have to do is wipe noses and take the kids to the toilet, and that's it, you know. I did have, I think for Karen's fourth birthday, a traditional party with games, and er, I'm not doing it again, because it's such hard work.

(mother of four children aged 3–8, two-parent, middle-class family)

As this mother explains, parties in commercial playgrounds mean less work for the adults. Other adults who are less motivated by convenience and who are prepared to undertake the extra work involved in organising a home party, are finding that their conception of a party does not match those of their children:

SOPHIE: They're used to things like Fun Zone now, so in a way I can't offer the same family party. They like to go climbing and running wild and things . . . They don't want [party games] now. They want something different . . . to run a bit wild and then sit down and eat, and they can't do that in a house really.

(mother of two children aged 5–7, two-parent middle-class family)

PAUL: I was going round saying, 'Let's have a game', and nobody came, so I just thought, 'Forget it', . . . I mean, it only lasted a couple of hours and then we had something to eat and the cake. Sang a couple of songs and that was the time up.

(father of two children aged 5–7, two-parent middle-class family)

An alternative response to the commercial playground is to commercialise the home party, and various services have been introduced in recent years, from all-inclusive party packages, hire of soft-play facilities and specialised party performers. At the same time as the home party is changing, the party in the commercial playground is attempting to sustain party traditions. Traditional party games, food and a party bag are considered an integral element of the package.

MR DOUGLAS: What we're saying is that we provide a party plate, it's party food, they get like Swiss roll, chocolate fingers, crisps, what you would normally get if you would lay out a party at home.

(Area Manager, chain of indoor soft-play centres)

KEVIN: And at the end, when like if you've been to a party or something, at the end, you get a present and they give you a present free. And after that erm, when like you know somebody and they thingy, they let you go in the back when it's your birthday and serve people with sweets.

(boy aged 6, working-class family)

SARA: Um, I liked it a lot because this bit right on top where you have your parties and big lunch boxes and there's this balloon slide and you . . . and you've got to go anywhere you like, and you go . . . bounce down it. It's really good fun. And you get it filmed 'cause somebody films your birthday and I've got erm two, three, no two ones in the living room.

(girl aged 8, middle-class family)

However, for some parents the commercial playground fails to provide some of the key ingredients of the birthday party, such as the exclusive focus on the birthday child in a defined space, or even interaction between children and family adults in play and games at parties. The prospect of hosting the private party in a busy collective play environment is therefore one that is not worthy of consideration for some. However, service providers are now increasingly concerned that it is the parents of the birthday child who are resistant to, or marginalised by, commercial party provision.

SALLY: To me it's not fair. And it's like Sean's party in erm January, and we're thinking 'where shall we go?' And unless we know that Kidz Kingdom's going to be empty I'm not going to go there, even though he loves it . . . I did go to one really good party and it was Amy's next door, and she booked the five to seven . . .

(mother of one child aged 4)

MR DOUGLAS: I say to them [the staff] 'Ask the parents if they want to play the game. They'll probably say no, but at least they've been asked.' Some of them do, some of them will get up and play the games with the kids. And that's the next stage now, it's to introduce those skills with the parents.

(Area Manager, chain of indoor soft-play centres)

Indeed, from the standpoint of the service provider, the birthday party presents an opportunity to establish personalised relations with children and adults who use their facility. Far from depersonalising the birthday party, it is considered that they are personalising the space of the play centre, and this is the message that they are attempting to convey to parents:

MR DOUGLAS: I *do* like children, I *love* children. It's only really in parties that we're that close with them, you know . . . When they come up here they have their own host . . . they [the host] would be with them for the half an hour or forty-five minutes that they're up [in the party room]. They make a fuss of the birthday child, they talk to the children, they play games with them.

(Area Manager, chain of indoor soft-play centres)

The children's birthday party contributes significantly to the changing geography of children's environments and adult–child relations. While the home has grown in importance as a site for children's play with the growth in ownership of electronic and computerised toys and games, clearly it would be misleading to suggest that only the home has opened up to children's play. At the same time, the birthday party has been drawn away from the home to the commercial playground. This in turn impacts upon the nature of commercial playgrounds: engaging children and parents directly in (party) activity is encouraged as an attempt is made to provide a personal space for the party.

The party's over

The geography of children's playspace is changing once more. Having contracted at the behest of parents and their fears for their children's safety, more expansive geographies are once again emerging through cyberspace, community provision and the development of commercial playgrounds for young children. The independent mobility which characterised children's lives of yesteryear has not, however, been recovered: parental presence is a prerequisite for participation in the majority of the commercial playgrounds described in this chapter. Such playgrounds feature prominently in the lives of young children. Much use is made of commercial playgrounds for birthday parties for pre-school-age children and children in the early years of schooling. Beyond this age, children tend to 'move on' to other commercial venues for the celebration of this annual event. Many parents welcome birthday parties in commercial playgrounds, the parties being perceived as a convenience by the majority and a necessity by a minority. Parties are not without their problems and children, parents and service providers expressed concern over the cost and the nature of the birthday package. Furthermore, exclusions on the grounds of (dis)abilities, cost, age and geo-social context have been identified. These misgivings strike at the heart of our concern over the place of these playgrounds in the landscape of children's play: Are they environments for children and, if so, are they environments for *all* children?

How far do the commercial playgrounds and the birthday party experience measure up to what might be conceived of as a *children's environment*? To date, this broader debate has focused on the question of segregation-by-default (Matthews 1992; Cunningham and Jones 1999). That is, concern is expressed that the provision of children's playspaces best serves the purposes of adults in that they are a means to restrict children's use of space by limiting use of the environment to designated areas. The Business of Children's Play project tends to suggest that this may be an unhelpful angle through which to engage commercial playgrounds with this wider debate. Undoubtedly, these centres serve a useful function for adults, undoubtedly centres pander to parents' often irrational concerns for children's safety, and much of the birthday party package is designed with adults in mind. Moreover, geo-social differences mean that such experiences are more available to some children than to others. However, it would be misleading to suggest that this new wave of development contributes only towards the geo-social exclusion of children in society. If anything, the shift in the birthday party from homespace to commercial playspace represents an extension of children's environments in society. Furthermore, and as has been noted elsewhere (McKendrick *et al.* 1999), commercial playgrounds are endowed with symbolic value in that they ascribe the right to play in domains and locales which were hitherto the preserve of adults. The playgrounds themselves harbour the potential to attend to the needs of the individual child at significant moments in her or his life, for example a child-centred approach to play is readily apparent at birthday parties.

More generally, there is also the issue of the links between the commercial playgrounds and other sites of play. The most significant points to be gleaned from this review of birthday parties in commercial playgrounds, concerns the changing position of homespace (Sibley 1995). The home is the most important everyday space for the majority of children. One of its key additional functions – as a forum for the celebration of the child through her or his birthday party – is being voluntarily eroded by homemakers with the greater availability of alternative venues in the commercial sector. Often accounted for in functional terms (saving time, cost and effort), this drift from the home is of deeper significance. The privatisation of homespace is exacerbated by this trend: the home no longer is the domain into which the wider community comes to celebrate the child. The celebrations of individuals with family/community occur in an altogether less personal domain. Ironically, it is the commercial play providers who perceive the need to 'introduce' to parents skills in party games and to sustain the rituals of 'happy birthday' and birthday cakes. Furthermore, it is the commercial playground where play opportunities for children are consciously provided. Far from striking the death knell for tradition, the new commercial playground may be a link-space which provides a lifeline to cultural customs such as those associated with children's birthday parties.

Notes

1 As with van Vliet's geographical framework, this paper reflects on children's environments in the *developed* world. In the context of the underdeveloped world, the workplace would be as, if not more, readily associated with children as the school, home and near neighbourhood.
2 For brevity, family pubs/restaurants with play areas are described as 'family pubs/restaurants' in this chapter: It should be recognised that not all family pubs/restaurants have play areas.
3 Throughout the paper, all personal, place and brand names in quotations have been given pseudonyms.

References

Alexander, J. (1993) 'The external costs of escorting children', in M. Hillman (ed.) *Children, Transport and Quality of Life*, London: PSI.

Ariès, P. (1962) *Centuries of Childhood*, London: Pimlico.

Armstrong, N. (1993) 'Independent mobility and children's physical development', in M. Hillman (ed.) *Children, Transport and Quality of Life*, London: PSI.

Barker, J. and Smith, F. (1999) 'From "Ninja Turtles" to the "Spice Girls": children's participation in the development of out-of-school play environments', *Built Environment* 25, 1: 35–43.

Bentley, T. (1998) *Learning Beyond the Classroom: Education for a Changing World*, London: Routledge.

Blakely, K. (1993) 'Parents' conceptions of social dangers to children in the urban environment', *Children's Environments* 11, 1: 16–25.

Cunningham, C.J. and Jones, M. (1999) 'The playground: a confession of failure?', *Built Environment* 25, 1: 11–17.

Davis, A. and Jones, L. (1996) 'The children's enclosure', *Town and Country Planning*, September, 233–5.

Furlong, A., Cartmel, F., Powney, J. and Hall, S. (1997) *Evaluating Youth Work with Vulnerable Young People*, Edinburgh: Scottish Council for Research in Education.

Gaster, S. (1991) 'Urban children's access to their neighbourhood: changes over three generations', *Environment and Behaviour* 23, 1: 70–85.

Hart, R. (1979) *Children's Experience of Place*, New York: Irvington.

Hillman, M. (ed.) (1993) *Children, Transport and Quality of Life*, London: PSI.

Hutton, W. (1995) 'The 30–30–40 society', *Regional Studies* 29, 8: 719–21.

Jeffs, T. and Smith, M.K. (1996) 'Getting the dirtbags off the streets', *Youth and Policy* 53: 1–14.

Kline, S. (1993) *Out of the Garden: Toys, TV and Children's Culture in the Age of Marketing*, London: Verso.

Matthews, M.H. (1987) 'Gender, home range and environmental cognition', *Transactions of the Institute of British Geographers* 12: 43–56.

Matthews, M.H. (1992) *Making Sense of Place: Children's Understanding of Large-Scale Environments*, London: Routledge.

McGallagly, J., Power, K., Littlewood, P. and Meikle, J. (1998) 'Evaluation of the Hamilton Child Safety Initiative', *Crime and Criminal Justice Research Findings Number 24*, Edinburgh: Scottish Office Central Research Unit.

McKendrick, J.H. (1999) 'Multi-method research', *Professional Geographer* 51, 1: 40–50.

McKendrick, J.H., Fielder, A.V. and Bradford, M.G. (1998) 'Commercial play centres for children,' *BoCP Project Paper 1*, Glasgow: Glasgow Caledonian University, 2–4.

McKendrick, J.H., Fielder, A.V. and Bradford, M.G. (1999) 'Privatisation of collective play spaces in the UK', *Built Environment* 25, 1: 44–57.

McNeal, J. (1992) *Kids as Customers: A Handbook of Marketing to Children*, New York: Lexington.

McNeish, D. and Roberts, H. (1995) *Playing it Safe: Today's Children at Play*, London: Barnardo's.

Moore, R.C. (1986) *Childhood's Domain: Play and Place in Childhood Development*, London: Croom Helm.

Nava, M. (1984) 'Youth service provision, social order and the question of girls', in A. McRobbie and M. Nava (eds) *Gender and Generation*, London: MacMillan.

Roberts, H., Smith, S.J. and Bryce, C. (1995) *Children at Risk: Safety as a Social Value*, Buckingham: Open University Press.

Sibley, D. (1995) 'Families and domestic routines: constructing the boundaries of childhood', in S.J. Pile and N. Thrift (eds) *Mapping the Subject: Geographies of Cultural Transformation*, London: Routledge.

South Ayrshire Council (1999) *The Space Place*, Ayr: South Ayrshire Council.

Valentine, G. (1996) 'Angels and devils: moral landscapes of childhood', *Environment and Planning D* 14: 581–99.

Valentine, G. and McKendrick, J.H. (1997) 'Children's outdoor play: exploring parental concerns about children's safety and the changing nature of childhood', *Geoforum* 28, 2: 219–35.

van Vliet, W. (1983) 'Exploring the fourth environment: an examination of the home range of city and suburban teenagers', *Environment and Behaviour* 15: 567–88.

Ward, C. (1979) *The Child in the City*, London: Architectural Press.

Webb, M. (1997) *Grounds for Play: A Report on Policy and Provision of Children's Playgrounds by Local Authorities*, Dublin: Institute of Technology.

Part 2

LIVING

7

PLAY, RIGHTS AND BORDERS

Gender-bound parents and the social construction of children

Stuart C. Aitken

Introduction

At a recent birthday party for a friend's 6-year-old hosted by The Discovery Zone, eight girls sat around a table eating pizza after spending their allotted time climbing, sliding and video-gaming in the facility's main room. An employee of the Zone asked my daughter what she would like to be when she grew up. 'I wanna be a kid' Catherine replied to smiles and 'oh how cute!' whispers from some of the adults present. This brief exchange has sat awkwardly with me for almost a year now, and sets a stage for three concerns that I discuss in this chapter. First, my daughter's response highlights a child's wilfulness in the construction of her own identity. That said, the somewhat patronising way the statement is encouraged by the audience of adults suggests a form of child–adult interaction which may be at odds with contemporary academic discourses focusing on children's empowerment. All children, from infants to youths, are wilful participants in the construction of their own identities, but those identities are also fomented through smiles, reassurances, tensions and aggressions of caregivers.

My second concern with the exchange is that it suggests an expression of caregivers' needs and desires as much as it does those of a child. My empirical focus in this essay is with the caregivers of pre-linguistic infants and, as such, I am particularly concerned with how infants' desires, needs and subjectivity are 'described' by adults. I also want to say something about how some academic feminists (including myself) construct parenting requirements, with the suggestion that those constructions may not necessarily promote the right response to infants' (or children's) needs.

Third, the exchange took place in a popular American game and pizza parlour for infants and young children which advertises itself as a place where 'a kid can be a kid'. For me, Catherine's response joins with the advertising of The Discovery Zone to raise questions about how the institutional environments of parenting and social reproduction are changing. Sometimes dubbed 'kid corrals' (Katz 1991) or 'branded spaces' (Davis 1997), commercial establishments such as The Discovery Zone are joined by the small entrepreneurial businesses and larger chains of private day-care centres such as Kindercare that now anchor many urban and suburban neighbourhoods. The disassociation of parents from increasingly younger children and the creation of institutionalised and commodified childhood experiences raises questions regarding the changing interdependencies between children and their parents.

My three concerns revolve around spaces of identity formation and are grounded by an interest in the gendered boundaries and borders that form between parents and infants. One larger question guides all three concerns: In what ways do institutions, parents and young children collide rather than connect in the pursuit of needs and control? A partial answer comes from my reading of object-relations theory, which today is construed as an attempt to understand how infants (and adults) bound and border themselves with and against the exegesis of an uncertain and often threatening outside world. To suggest that the theory assumes an internalisation of the outside world as objects is a gross simplification because it elides the processes, joys, fears, social constructions, transformations and transitions that comprise a search for identity. Aspects of the body of work that surrounds the theory are liberatory because they do not necessarily suggest the outcome of a child's development, but I am nonetheless concerned by some of the infant–parent constructions that are assumed. I argue that although object-relations theory enables a post-structural focus on the rela-tions between parents and young children, that focus may be somewhat myopic.

The chapter provides a second answer to my larger question, in part through excerpts from protracted discussions with parents and primary caregivers in San Diego.[1] Our conversations began during pregnancy and continued until the child was 3 years old (cf. Aitken 1998: 1–30). They provided qualitative infor-mation on specific day-to-day events, household members' feelings, attitudes and interdependencies, from which I discuss child–parent borders and gender identity formation. I also use these conversations to engage larger concerns with how childhood is institutionalised through a specific focus on childcare and parent–child identity politics.

The architecture of the chapter is as follows. I begin by elaborating on the corpus of knowledge that has come to be known as object-relations theory and suggest why some of the new work in this area makes me uneasy. I then note the implications of considering young infants as wilful beings rather than as passive dependants. Although I try not to lose sight of the larger theoretical implications surrounding the paper, what follows is grounded in the day-to-day experiences of a small selection of young parents and my interpretations of those experiences. I close the chapter by elaborating on the ways that self-esteem and well-being may be facilitated or frustrated through institutionalised day-care and what this means for the rights of children.

Playing with identities is a dangerous game

Recent feminist and post-structural accounts of 'personal geographies' and constructions of self are profoundly shaped by the work of object-relations theorists Melanie Klein and Donald Winnicott (cf. Aitken and Herman 1997; Bondi 1999; Kirby 1996; Robins 1996; Sibley 1995a, 1995b). Developed as an attempt to revise regressive and individualistic aspects of Freudian psychoanalytic theory, the object-relations of Klein and Winnicott render the self as more social and playful. Their depiction of object-relations is richer than those developed by Freud because other subjects are not converted into objects (e.g. mothers) and the recalcitrant material of narcissistic drives. Klein is concerned with the development of the social self. She suggests that infantile fantasies – many of which are violent, sadistic and paranoid – are associated with discomfort. Countering Freud, Klein notes that mothers are not responsible for infantile fantasies or the emotions of their children. Rather, the infant's earliest experience of social relationships is when a caregiver – usually the mother – provides comfort against hunger, cold, and so forth. Moreover, any pre-Oedipal oneness with the mother is lost when the child develops a sense of borders and selfhood, and a sense of the social (Sibley 1995a: 6).

Winnicott, whose psychiatric practice was supervised by Klein for three years, argues that the whole social field is held together in a transitional space, relating subjects to each other and to reality while at the same time maintaining a certain amount of slack and flexibility. Transitional spaces are for 'play' and reconfiguration, belonging neither to the subject nor to some existent reality. In Winnicott's formulation, an infant is able to experiment with her culture and environment, and the nature of 'childhood' and culture change as the objects around the child (such as social relations and family structures) change. The child not only 'becomes' through the influence of cultural, social and political environments but she also brings something of herself 'into creative living and into the whole [of] cultural life' (Winnicott 1971: 102). Common to both Klein's and Winnicott's work is the idea of a developing child through simultaneous inward and outward representations, and an emerging sense of borders and self, both real and imagined, as social and cultural constructions.

Nancy Chodorow (1974, 1978) uses object-relations theory to influence feminist discourse by suggesting that because a child's first internalised 'objects' most often are the breast or the mother, it may be possible to envisage a more liberatory focus on mother–daughter relations. Freudian psychological theory posits mothers as 'other', the dark continent and forbidden place of the Oedipal myth. Noting this focus on women as 'other', Chodorow points out that the maternal is almost completely missing from his account. She argues against Freud by suggesting that on a deep emotional level mothers experience daughters as less separate and individuated from themselves than sons.

Julia Kristeva attempts to answer questions about borders by focusing on the confusion and anxiety for a subject attempting to apprehend autonomy when faced with corporeal conduits between the internal and the external (e.g. nose, mouth, anus, penis, urethra, vagina). She argues that aversion to bodily excretions is a social construct which is metaphorically and linguistically linked to symbolic aversions to the 'other' that are embodied in racism and sexism. Kristeva (1982: 32) uses Winnicott's notion of the 'spaces of play' (transitional spaces) as provisional boundaries that help us understand our relations with the 'other'. In her rendering, transitional spaces are partial, containing semi-objects that are not quite real and not quite part of the subject. Kristeva's writing on 'abjection' (hopeless anxiety because of something we do not like but cannot get rid of) supplements Winnicott's ideas with an emphasis on textuality and the role of language in undergirding personalities and our placing of the 'other'. David Sibley (1995a: 8) draws on Klein and Kristeva to suggest that the ways some people construct themselves against a 'generalised other' is abject in the sense that there is an urge to make a separation, but this creates anxiety because such separations can never be achieved. Sibley's larger project suggests that abjection helps us to understand the geographies of exclusion that surround sexist and racist stereotyping.

Jane Flax (1990, 1993) is also concerned with the construction of difference. Her project is to transform Winnicott's notion of transitional spaces into ideas of justice that are not constrained by hierarchical and arbitrary valuations of difference (Aitken and Herman 1997: 83). Flax (1990: 116) argues that Winnicott's 'spaces of play' suggests one of the most important contributions to post-Enlightenment thinking because it de-centres reason and logic in favour of 'playing with' and 'making use of' as the qualities most characteristic of how the self develops. She reworks Winnicott's ideas into an approach to justice that jointly applies feminist concerns for the diminution of relations of dominance and post-structural concerns for the play of differences.

It seems clear that theorising about object relations helps some feminists and post-structuralists critique oppressive patriarchal systems. I am nonetheless troubled by assumptions made by Winnicott, Klein, Chodorow and others that influence how we are beginning to understand subjectivities and, specifically, those that develop around parent–infant relations. For example, I argue elsewhere that Winnicott's ideas are problematic because of their normative assumptions about mothering (Aitken and Herman 1997). In particular,

Winnicott stresses the centrality of mother–child relations in his rather annoying concept of the 'good-enough mother'. By suggesting that mothers can be 'good-enough mothers', Winnicott (1971: 11–13) refers to the 'illusion' of a possible reality created through the mother. As I understand his use of this concept, Winnicott is suggesting that by being good enough, the mother stands in for what the child creates in her imagination (hence his use of the term 'illusion'). The good-enough mother creates illusions that are also transitional objects which enable the child to play with, and reconfigure, her experiences without threat or challenge. Winnicott emphasises that the transitional object is safe and neutral because no-one asks the question: 'Did you conceive of this or was it presented to you from without?' After making this quite startling point that would seemingly draw us away from the patriarchal strictures of Freudian psychology, Winnicott then goes on to emphasise that once the illusion is in place (e.g. the breast as a transitional object), the main task of the mother is disillusionment (e.g. weaning). Although some feminist writers influenced by object-relations theory may note the essentialism embedded in this problem and its example, they usually side-step the issue and move on to what they consider to be Winnicott's post-Freudian contributions (cf. Flax 1993; Henriques et al. 1984). I argue that the way Winnicott highlights a precise structural dialectic (illusion then disillusion) as good enough for certain developmental outcomes in children is problematic. It does not help mothers to distance themselves from a set of binding and perhaps stultifying moral obligations around the development of their children. It is worrisome to me that object-relations theorists do not engage the morality of these obligations or, in a larger sense, how certain myths of parenting and child development are established and sustained.

Until quite recently, parenting strategies as they relate to stages of childhood (e.g. weaning and potty-training) were often assumed to be unproblematic in the child development and geography literature that deals with childcare. Child development studies often implicitly adopted a unidirectional approach wherein focus is on how adults direct and control interactions, while there is a conspicuous lack of interest in the infant's contributions (Hoogsteder, Maier and Elbers 1998). The geography literature usually focuses on either institutional structures or the agency of mothers (cf. Dyck 1990; Holloway 1998a, 1998b; Rose 1993). For example, Kim England's (1996) edited volume on geographies of childcare and working mothers provides some excellent examples of institutional support and how childcare provision fails mothers, but the needs of infants and children are never made explicit or problematised. These studies assume that parents, for the most part, love their children and will do the best they can to provide for a set of universally understood needs in the face of oppressive societal structures. Sarah Holloway (1998a) is one of the first feminist geographers to question this universality in terms of what counts as childcare and who, specifically, needs it. By so doing, she begins the difficult task of problematising the relations between parents and childcare institutions and engages critically how childrearing is constituted. Holloway's work dovetails with that of Isabel Dyck (1996) because it focuses on childcare cultures. Dyck (1996: 126) argues that how women define and negotiate a space for childcare is not merely 'local' but revolves around the complex changing daily lives of women (and men). Holloway (1998b: 31) recognises that a 'moral geography of mothering' is constituted in a 'localized discourse concerned with what is considered right and wrong in the raising of children'. I want to take Holloway's point further by suggesting that although the day-to-day *childcare* is defined by local discourses, the larger *needs of the children* are often defined by universal agendas that suggest certain kinds of development. Kenneth Gergen and his colleagues (1990), for example, note that the majority of German and American mothers in their cross-cultural study recognise linear stages of development in their children. The stages are used in relatively unquestioned ways by many educational and childcare institutions. Kindercare childcare centres in the US and the UK are structured into rooms and activities that segment children by ages that are suggestive of specific Piagetian learning stages.

Even someone as concerned about social contexts as Nancy Chodorow seems to impose a fixed meaning on infantile consciousness upon which essentialist and universal notions of gender are mapped. In her account, mothers' treatment of their daughters makes them both more dependent and more expressive than males. Chodorow argues that the mother's internalisation of sexual difference creates sexual difference in the child. Eleanor Maccoby (1988, 1990) contests the essentialism in this assumption with the suggestion that both boys and girls learn nurturing and maternal roles early on, but boys are later constrained to deny this capacity. Clearly, there are still universal assumptions in Maccoby's indictment of Chodorow's work. I want to argue with Janice Doane and Devon Hodges (1992) that it might be more productive to recognise the extent to which our internalisation of idealised and mythic parenting requirements is at work to preserve gender distinctions. I have written elsewhere about the continued hegemony of a mythic nuclear family norm that has very little to do with the work of parenting (Aitken 1998, 1999). The point that Doan and Hodges raise, and the argument I want to make here, relates to our need to imagine that these parenting requirements do not necessarily promote the right response to infants' needs:

> The object-relations emphasis on early infantile experiences tends to de-emphasise the role of culture and representation in constructing that otherwise may appear to be pre-linguistic entities such as the preoedipal period, infantile fantasy, and, often, gender itself.
>
> (Doan and Hodges 1992: 2)

124

I argue that object-relations theorists, and those influenced by them, not only forget mythic representations and culture, but also give short shrift to the possibilities of unique geographic and historical contexts. The post-structural refocusing of spatial justice in the work of Dyck (1996) and Holloway (1998a, 1998b) moves us closer to understanding childcare cultures in terms of these important contexts, but we remain stuck in reified patriarchal notions of child development (and child–parent relations) if we cannot excise childcare practices from fixed and universal ways of knowing infants.

Liz Bondi (1999: 18) pulls on Klein, Winnicott and others to suggest that the way psychoanalytic theory is sometimes used, 'within and beyond geography, arises from a sense that meanings or interpretations of subjectivity are somehow imposed'. With Bondi, I assume that although a merging of psychoanalysis and other social sciences is fruitful, in suggesting the delimitations of borders and boundaries it is often easy to incorporate subjectivities that leave little room for meanings generated by 'others' (in this case, parents and infants). It seems to me that object-relations theorists may be left with little to say if we suggest that the need for 'good-enough' parenting capabilities is not universal and that playing with illusion and disillusion (in the Winnicottian sense) is a dangerous game when constituted as natural (as Klein does). Suppose we dispense with the naturalness of the 'good-enough' mother (or father). Suppose, rather, we interrogate societal demands for parents' selflessness and, by so doing, question some of the unreachable myths of parenthood. Suppose, also, we let parents 'describe' their infants in ways that integrate their own needs and frustrations about work and relationships with their partners. With this joint strategy, I argue that we are better placed to understand how a child's 'being' may be at odds with adultist and institutional conceptions of what a child should be 'becoming' (James, Jenks and Prout 1998) and we move further towards understanding not only child-rearing cultures but the ways they constrain and contextualise children.

Bounding the self and bonding with children

In this section, I introduce some of the struggles parents engage with in attempt to maintain goals, borders and boundaries as they relate to their own self-identity and that of their children. I begin with examples that suggest mythic parenting roles. I then focus on parents' attempts to negotiate time and reconfigure space on a day-to-day-basis and how those attempts may foment gendered identities.

The essentials of natural parenting

The complexity of child-rearing often stimulates the kinds of changes that highlight the structural fragility of contemporary families and, as I argue elsewhere, the parenting norms of motherhood and fatherhood are often nothing more than misplaced attempts to bind the family as a functioning unit (Aitken 1998). Mythic renderings of parenthood and childhood foster political identity formation and social placement, but they do not reflect the day-to-day work of parenting, nor do they anticipate adequately the complex changes of daily living that accompany child-rearing. I have come to believe that the problem lies with not only how we represent parenthood, but also with how we categorise childhood. As dependants, infants are often seen as relatively passive appendages and even (if we are to believe some advertisements) as fashion accessories. It is important to recognise the necessity of infants' dependence on adults, but it is also important to recognise infants' negotiation and play around a transitional space and time that helps them establish 'selves' that are not necessarily dependent on 'good-enough' parenting. Infants are not passive dependants but wilful participants in family life and as such, an object-relations perspective that moves beyond the practice of parenting as illusion and disillusion can point to the purposeful and problematic relations that develop between caregivers and children:

> It's just more of an introspective thing with him [the baby], and when I'm with him, I'm *really* with him. It's very natural. I'm not somewhere else and I can just put my things aside.
>
> (Kathy)

> The baby, you have to be with him. I can't work [at home] . . . I can do some but sometimes I find when I'm with him I should play with him, lie down and just hang out with him. He's got to learn to entertain himself. But he does a lot of times. With me especially he's much better at that than with his Mom. When she's there, he groans and wants to be with her. I feel freer than my wife . . . When it's me he knows it's Dad and groaning and crying doesn't work.
>
> (Frank)

These two interview excerpts hint at 'essentialist' distinctions between a young mother's and a young father's relations with their respective infants. A 'natural' bond is suggested between the young mother and her child, whereas a more distanced relationship is suggested by the father's words.[2] These excerpts play masterfully into Klein's belief in 'natural' sexual difference and Chodorow's belief that women are less individuated than men. The literal 'meanings' of these excerpts notwithstanding, I am more concerned here with how positions such as these interplay with day-to-day contexts and the needs of parents and children. Although many of the parents in the San Diego study thought of motherhood and fatherhood as 'natural' gendered categories that gave meaning to their emotions and behaviours, further scrutiny suggests that these emotions and behaviours are complexly woven, unravelled and sometimes detached from the work of parenting. For example, the following excerpt from an interview with another young mother suggests something of Klein's infantile fantasies, but it also hints at distinctions in child-rearing practices between a mother and father:

> Wesley doesn't really have tantrums but he'll definitely let you know if he doesn't want to do something. But you can coax him out of it really easy. He'll get mad at something, like he loves to crawl up on the couch in the den and play with the phone and if I take it away from him he'll let me know he's mad. And he's *strong*. I'm pretty much the only one he does it to, but he'll bite me. If he's going to bite someone he'll nip me. I spank him on his diaper. I'm sure the amount I hit him, my husband just laughs at me 'do you think that hurt him?' But he'll get a pouted lip and he'll know I'm mad. He hasn't done the throwing himself down on the ground, the kicking and the screaming, yet!
>
> (Elaine)

Klein's counsel that, for infants, good and bad emanate from their mothers is born out by this excerpt, but there is also a suggestion here of the complex social environment that precipitates the gendering of child-rearing. The problem with Klein's suppositions is that they are embedded with a belief in the inevitability of a child assuming the 'correct' sexual orientation (Doane and Hodges 1992: 3). Alternatively, Judith Butler (1990, 1993) points out that gender is a performance, an illusion created by particular events in space and time which are repeated to the extent that they become immutable and mythic. The problem, in Winnicottian terms, is that the illusion is never disillusioned. And so, for Butler (1993: 2), gender is a 're-iterative and citational practice by which discourse produces the effects that it names'. The young father whose words are quoted earlier did coddle and nurture his daughter during our interview, but he had difficulty speaking of the act, much less discussing the emotions around it. His vocal account is part of a set of preferred gender norms that belie his actions. I will say more about how fathers may articulate and act out gender in different ways in a moment, but for now I want to think about the construction of identity boundaries.

If we raise young infants from the status of mere dependants with universal needs by recognising that they are unique, then we need to recognise also the implications of parents' gendered performances on the contexts of young children. At an early age, sexual, racial and class boundaries are constructed and maintained by children and caregivers. If these boundaries are ascribed as 'natural', then young children may not be allowed to play with their identities because they have to 'get it right' (Yelland and Grieshaber 1998: 2).

127

Controlling children and playing with boundaries

Playing with boundaries, controlling events and exploring ambiguities are important aspects of how we construct ourselves. Infants and parents continually dance around events as they try to define comfortable contexts for their developing relationship. Even when traditional gender roles within a nuclear family are reversed, conversations with some San Diego parents suggest that patriarchal social imaginaries are, for the most part, maintained.

Allen is a self-declared house-husband whose wife Janet maintained full-time employment through the four years we knew them.[3] I elaborate on his story here to suggest that changed gendered identities are rarely articulated in heterosexual relationships even though gender roles are played with and changed. Prior to the birth of their first child, Allen and Janet bought a small bungalow nestled in the lee of Caper Mountain, a county park within San Diego's metropolitan area. For most of his working life, Allen was employed in construction and the 'working-class feel of the neighbourhood' suited him well. Allen was forced to give up his fairly lucrative construction work with the downturn of the Southern Californian economy in the early 1990s. Instead, he began working with special education students at a local community college and was hoping to go back to school to obtain a credential to follow a career in that area. Just after Hannah was born, he decided to put his career on hold so that he could be his daughter's primary caregiver.

> [Hannah] is so good with everything. I haven't really seen her complain about anything, except when she's tired. She'll go back in her room and point to her bed and I'll put her down. The only real trick is that if I really have to have something done and she needs to be with me, then I'll try to wear her out early and get a nap in before I have to go do what I have to do with her.
>
> (Allen)

This excerpt from our first set of interviews suggests that although Allen perceives that he has a good relationship with his daughter, there are certain games he will play to tire her out. Two difficult questions are raised by Allen's concerns. What constitutes control and how is a parent's or child's sense-of-self moulded through enabling, deceiving or cajoling? In what sense are parents' day-to-day needs mapped onto their relations with their children? To help answer these questions, I want to compare Allen's situation with that of Maria. I also want to highlight some similarities in the ways these two parents connect with their children. Maria lives in an urban community eight miles to the west of San Diego in a house that she and her husband Bobby bought just after Josephine was born.

Bobby is in the Navy and during the period of our interviews was stationed in Seattle with shore-leave to visit his family one weekend out of every month. This situation left Maria to deal with most of the parenting responsibilities which, she told us, really frustrated her career aspirations.

As with Allen, Maria's control issues and the borders between herself and her child are often fomented in games and in play. Here's what Maria had to say about cajoling and playing to persuade a recalcitrant child:

> That is probably my biggest time problem right now, is when she gets up in the morning, she doesn't want to wake up, she doesn't want to get dressed, she wants to do everything herself. So it's a big struggle! But if you let her pretend that she's doing it herself, she's actually in a much better mood.
>
> (Maria)

Maria plays a game of giving her daughter autonomy over dressing herself, but Josephine will often act out a defiant sense-of-self which is, as Maria puts it, 'another story'. Stalling tactics are an issue for many parents to the extent that popular self-help books elaborate on strategies to placate children or bend them to the will of the caregiver.

Just over a year, and another child, later, Allen's sense of control over domestic responsibilities is conflated with what he perceives as the 'naturalness' of his parenting:

> [The children] are pretty good at knowing when it's meal time. You have to feed them, get them naps and that kind of stuff, so it's, you know, a *natural* schedule as opposed to an *unnatural* one where you force yourself to start working at five in the morning digging a hole or whatever in construction.
>
> (Allen, *my emphasis*)

Using the rhythms of the day and his children's needs for food, nurturing and recreation enables Allen to map out appropriate parenting responses from his own ideas of what the work of fathering entails. Alternatively, Maria's work, career and child-rearing responsibilities reflect a frantic schedule that is quite different from Allen's seemingly serene daily rhythms. One year later, a more serious sense of frustration had crept into her discussion of her relationship with Josephine:

> But if you're short of time, forget it. And the same thing with going to bed: she doesn't want to go to bed; she wants another story, another

song; she stretches it out. It's not the normal activities . . . it's just the fact that she's tired, wants too much control and wants to do every-thing herself.

(Maria)

This spatial and temporal partitioning may be a coping method, but it may also reflect a pernicious geography of exclusion (Sibley 1995b). Sibley (1995b: 93) argues that obsessive control of space and time by parents leads to exclusionary practices that may have some bearing on later behavioural problems in children. Wood and Beck (1990, 1994) point out that the space of the home is a conduit for rules and regulations which demarcate appropriate behaviours in children. The important point about this work for what I want to say here is that those rules and regulations are tied to larger-scale societal and global influences. The space of the home and the time of family members are contested grounds and locuses of power relations that go beyond individual family members.

Parents needs vs. commodified children

David Oldman (1994) created the term 'childwork' to describe professional childcare as a category of capitalism that exploits the value of children's growth and development. For example, Kindercare Learning Centres managed more than 1,200 neighbourhood facilities in North America and Great Britain in 1997. Although Kindercare is the single largest private provider of childcare in the United States, it still controls less than 1 per cent of the national market and is placing itself for substantial expansion in the next decade (Davis 1997). Commercial day-care centres join with The Discovery Zone and Chuck E. Cheese's as common Southern Californian havens for 'child care' in the midst of an urban environment that is perceived by many parents to be threatening and hostile. While academics ponder the unfathomable complexity of contem-porary families, these entrepreneurial businesses are seizing upon a niche of opportunities for investment. In the following pages, I tentatively relate the ways interdependencies between children and their parents are extenuated through the spatial segregation of day-care and the segmentation of social relations around gender, class and age.

I focus on two teachers for whom childwork respectively facilitates and frus-trates their mothering efforts. Sophia lives in a detached single-family house with a view of the ocean. Her husband is in the merchant navy and spends seventy-five days at a time on the sea followed by seventy-five days of shore-leave. It seems that Sophia is responsible for domestic work even when her husband is home: 'he has his own projects around the house that keep him busy'. Sophia works in a school in La Mesa, a community ten miles to the west of where she lives.

This first extended excerpt from Sophia suggests the inordinate amount of energy spent on selecting professional childcare, the difficulty of finding care to suite specific needs, and the race and class biases that sometimes accompany childcare decisions:

> He's been with a sitter since he was six weeks old and he's never cried when I've left, never. He wore out a new sitter last year then we had a different one who was wonderful but she moved. So I had to find a new one this summer. This was kind of stressful but these [new] guys are great . . . I went to a service and talked to about twenty-five people on the phone and made all sorts of appointments. Either they were cold or we didn't meet their qualifications: they didn't take them until maybe 7:20 a.m. or – it sounds horrible – they had a distinct accent and I didn't even understand them, . . . or they didn't take infants and I was going to bring a new-born. There was one that was a couple of blocks over, then she told me where she lived, in an apartment. I was like "oh my god, I don't want him in an apartment all day." So there was these women who had just started up and they wanted to cater to teachers, they want to be on a school schedule, and we don't have to pay on our holidays (while others) do. And they're great.
>
> (Sophia)

Sophia's unwillingness to leave Denny in an apartment or with people who had distinct accents signals exclusionary tactics that speak to issues of urban segmentation through ethnicity and space. The space of an apartment may indeed be inappropriate for some children, but it is often the only option that many people can afford in day-care. As commercial day-care centres grow to fill certain market niches, it is clear that the market is divided into temporal and spatial segments. The more selective a parent is, the more costly the day-care. Sophia's compromise was to find something outside of the neighbourhood that fit her schedule as a teacher. This second excerpt, from Sophia's next interview, suggests implicit concerns for Denny's welfare as well as the gratitude of a working mother coming home to a tired infant:

> I thought he might be whiny or be one of those 'miss you mommy' when I leave, but the place I take him to, they have so many toys and so many different toys from what he's got. They have the bigger toys, like a little kitchen table, the play set, all the little play school stuff and the little cars. He loves it. And he's playing so much harder now; we're back to him easily going to sleep at 7:30. I can tell he's tired because

he's playing all day. I came over one day [to sneak a peak] and he was in their little pool, happy as a clam.

(Sophia)

Attending to how Sophia describes Denny's sense of well-being, it is difficult not to focus on her own feelings of relief at picking up a tired child at the end of her own hectic day. Although it is not my intent to belittle the needs of parents or to suggest that parents should spend more time with their children, it is nonetheless important to recognise that the needs of children are described in close association with the needs of parents. That this should be the case with pre-linguistic children does not detract from my point that we need to focus careful attention on how parents' 'describe' their infant's development. I believe that this is where object-relations theorists fail to appreciate the importance of geographical and cultural contexts.

A different context is highlighted by Alishia, a single parent who works as an outdoor science teacher for children in Sixth Grade. During the school year she takes classes on field trips and during the summer she organises a summer camp in the mountains near San Diego. Her income is based on the needs of specific schools for science field-trips and those needs are quite variable. Alishia does not know how much work she will get in any particular month and sometimes she is required to drive considerable distances to get to different schools in San Diego County. Her stipend for running the summer camp is a necessary supplement to a meagre income. Alishia's context highlights how many of the issues already discussed in this chapter – juggling career and childcare, and the lack of convenient facilities to help alleviate the complexities of day-to-day life with a young infant – are exacerbated for low-income single mothers. Identity issues, in particular, collapse in a quagmire of stress and anxiety that is reflected in Alishia's sense-of-self and, as time goes on, also in aspects of her relationship with her son Alex.

When pregnant, Alishia was particularly anxious that no 24-hour baby service was available for someone of her income and frustrated over the lack of any structural support for women like herself:

I work 24 hour shifts so I'm really concerned about what I'm going to do when I work those shifts with my baby. So day-care, 24-hour day-care; actually, 24-hour everything. I do a lot of things at night, or always have in the past, and because I live close to the main Post Office, and lots of stores are open 24-hours, I've been able to do . . . That is how I avoid crowds. Day-care, that's what is concerning me now, mostly. Actually, I really need *nightcare* [laughs].

(Alishia)

A year later, Alex was eight months old and, at least during the interview, a demanding and precocious youngster. On several occasions, Alex screamed until Alishia paid attention. Once again, crucial concerns with day-care dominated our conversation:

> After I pick him up from day-care [between 6 and 7 p.m.] is the worst time . . . you know the baby-sitters tend to have them nap as much as possible, at the wrong times, so when you get them they're tired and they're hungry, but you don't want to put them to bed yet otherwise they won't sleep. *Just a cranky hour.*
>
> I work for the County Board of Education and they're still not very understanding about having a childcare facility. I work in the mountains in the summer. They don't even let us take our kids up, you know, like when I was a camp director this summer . . . and there really wasn't any reason why he couldn't have been up there. *It seems kind of anti-child* [Alex screaming].
>
> (Alishia)

Our third interview with Alishia, just before Alex's second birthday, provided a clear sense of the boundary between her work, her self and her child. She pointed out that the childcare facility had '*terminated me*' because of disputes about the terms of their original agreement. Alex was at the facility – a small commercial day-care business – for two years. Alishia said that the woman who ran it was originally very flexible and aimed to please the parents in terms of the way she cared for the children. As her business expanded and she built up a waiting list for the service, the terms of care changed and the woman became inflexible in her practices, insisting that she decide what the children eat and when they nap. Apparently, this was to give the woman a block of time to clean house and make phone calls. This kind of structure might be appropriate for a large day-care facility but, as far as Alishia was concerned, it was inappropriate for a small neighbourhood business. The woman gave Alishia 'a take it or leave it attitude' which 'forced' her to withdraw. At the same time, Alishia was loath to break Alex's attachment to the other kids at the day-care service as well as disrupt his (and her) routine. She was hoping to shift him to a pre-school and then keep him there until he started public school. This was a source of concern for Alishia because most pre-schools in which she would ideally like to place him had long waiting lists.

Day-care is important. Constraints on access to public facilities and support often result in families struggling privately and imploding in upon themselves. There is a politics of difference here that incorporates not only issues of distribution and moral geographies of mothering but also complex interdependencies between mothers' needs, children's needs, work schedules and habits, career opportunities, personal demeanours and external constraints. Whereas Sophia lucked into something she could just about afford with opening times tailored to her profession (albeit in a somewhat inconvenient location), Alishia spent three frustrating years without finding a satisfactory solution. In the absence of a broad public discussion on the importance of state-sponsored day-care, mothers like Alishia often bend to the whim of private centres. I do not want to suggest that the relationship between Alishia and Alex was strained by this context or that it coloured his sense-of-self. Such an assessment is well beyond the confines of this study. Nonetheless, there is clearly tension in their lives, and during interviews Alishia continually referenced personal problems and stress with issues related to finding adequate day-care. What she considers appropriate for Alex may be unattainable from her local childcare culture. The childcare contexts of Alishia and Sophia leads me to some concluding thoughts on the rights of infants and children, and the ways that those rights are abused by the restructuring of childcare landscapes.

The rights of children and re-thinking the ties that bind

In the absence of public discussion in the US on the importance of state-sponsored day-care, long waiting lists are common for most commercial day-care centres, particularly 'respectable' chains like Kindercare. Cindi Katz (1993) and Susan Davis (1997) argue that these enterprises are a shameless commodification of children's lives, designed and promoted on the basis of fear (of unsafe streets, unsupervised youths and youth of colour) and media 'terror tales' (of abduction and molestation by strangers). My interests go beyond the important structural concerns of Katz and Davis to the argument that if private commercial day-care centres can play on parents' fears and anxieties, then we are inadequately conceptualising infants' and children's spatial rights. In a recent book, Sylvia Anne Hewlett and Cornel West (1998: 28) argue that the basic components of good parenting – caring, nurturing and cherishing – are no longer adequately supported in our society. They point out that this is primarily because good parenting comprises a set of values and activities with limited market potential. But good parenting, like the Winnicottian spectre of the 'good-enough' mother, is not the central issue on which we need to focus. Rather, we need to interrogate the selflessness demanded of parental requirements and focus on mythic representations of parenting ideals so that we are better placed to understand how a child's rights and 'being' may be at odds with conceptions of what adults think a child should be 'becoming'. Jane Flax (1993: 112) argues that our political life and, I would add, our well-being, is partially constituted through and necessitated by differences and interdependencies. Segmented day-care land-scapes erode this well-being as surely as walled and gated communities. We must be suspicious of a spatial justice based on the logic of segmented markets because it hinges on an understanding of space and children that is essentialist, exclusive and controlling. By appealing to the rights of children for accessible and affordable day-care, we need to problematise more fully what we mean by those rights, because sometimes we appeal also to ongoing forms of domination that exist in the construction of gender, race, ethnicity and class.

And so I return to my daughter's adamant assertion of self at The Discovery Zone, so that she may not be discounted as a monadic and volitional person and so that we understand more fully the implication of the adult 'oh how cute' responses as a form of 'describing' childhood. Gill Valentine (1997a, 1997b) points out that childhood and parenthood are cultural inventions and ideologies that are (re)constructed and (re)produced over time. That family contexts are changing is quite clear, and it is indisputable that today's parents are managing their lives and those of their children in increasingly complex ways, but the effects of these changes on the identity relations of young children and parents are far from clear. I am concerned that academic literature rarely considers the complex relations between parents and infants, how parents describe children, or what it is to be a child. Valentine (1997c) notes that children are often imag-ined as vulnerable, incompetent and in need of protection. Because children are cast as innocents, parents bare the brunt of the failed reconfiguration of insti-tutions that support social reproduction. For working parents, the complex web of childcare is often reduced to finding whatever is accessible and affordable. The personal geographies of parents and children are then reconfigured (usually with some stress) to suite the timing and spacing of day-care and entertainment facilities. Chains like The Discovery Zone free parents from intense childcare while formal supervision in day-care centres like Kindercare stimulates a market for childwork. Children's activities are increasingly structured around the economic interests of childwork, and decreased parent–child interaction increases the proportion of labour traded in the market. The commodification of chil-dren's lives in this and other ways is disconcerting because the complex and rich positive relations that develop between children and their parents and, ulti-mately, children's identities are stretched tightly around the needs of capital.

Acknowledgements

Research for this chapter was supported in part by grant SES-9113062 and SES 97–32469 from the National Science Foundation. Special thanks go to all the students who were employed through these grants. In addition, I would like to thank the families in San Diego who filled out our questionnaires or agreed to be interviewed as part of the study that drives large parts of the arguments in this chapter. Opinions, findings, and conclusions expressed in this chapter are mine and do not necessarily reflect the views of the National Science Foundation, San Diego State University, or the students and families involved in this project.

Notes

1   A larger study, of which this chapter is a part, comprised a geographically dispersed mail questionnaire of 577 adult members of the households within which the woman was expecting a child, and two follow-up surveys prior to the child's first and third birthdays. Because the 'gatekeeper' who provided access to the family was always a mother, there are no single fathers in the study. Women were contacted through a large Health Maintenance Organisation that caters primarily to low- and middle-income patients. There were no openly gay or lesbian parents in the study. All three surveys focus upon household members' day-to-day activities, responsibilities, and opinions about work, day-care and family life. The follow-up surveys enable an assessment of changes in gender roles and relations after the birth of the child. From this sample population, a smaller set of 127 in-depth interviews from adults in households expecting a first child was gathered. It is on these in-depth interviews that the empirical parts of this chapter are based.

2   'Essentialism' maps exclusive characteristics onto men and women, and 'naturalism' maintains that such qualities are genetic and biologically based rather than socially constructed.

3   Allen is introduced in Aitken (1998: 62–5) to suggest different constructions of fatherhood and then discussed again in Aitken (1999) as an example of the 'work of fathering'.

References

Aitken, S.C. (1998) *Family Fantasies and Community Space*, New Brunswick, NJ: Rutgers University Press.

—— (1999) 'Putting parents in their place: child-rearing rites and gender politics', in E. Teather (ed.) *Embodied Geographies and Rites of Passage*, London and New York: Routledge.

Aitken, S.C. and Herman T. (1997) 'Gender, power and crib geography: from transitional spaces to potential places', *Gender, Place & Culture: A Journal of Feminist Geography* 4, 1: 63–88.

Bondi, L. (1999) 'Stages on journeys: some remarks about human geography and psychotherapeutic practice', *Professional Geographer* 51, 1: 11–24.

Butler, J. (1990) *Gender Trouble: Feminism and the Subversion of Identity*, New York: Routledge.

—— (1993) *Bodies that Matter: On the Discursive Limits of "Sex"*, New York: Routledge.

Chodorow, N. (1974) 'Family structure and feminine personality', in M. Zimbalist, M. Rosaldo and L. Lamphere (eds) *Women, Culture and Society*, Stanford, CA: Stanford University Press.

—— (1978) *The Reproduction of Mothering*, Berkeley, CA: University of California Press.

Davis, S.G. (1997) 'Space jam: family values in the entertainment city', paper presented at the American Studies Annual Meeting.

Doane, J. and Hodges, D. (1992) 'From Klein to Kristeva: psychoanalytic feminism and the search for the 'good enough' mother', *From Klein to Kristeva: Psychoanalytic Feminism and the Search for the 'Good Enough' Mother*, Ann Arbor, MI: University of Michigan Press.

Dyck, I. (1990) 'Space, time and renegotiating motherhood: an exploration of the domestic workplace', *Environment and Planning D: Society and Space* 8: 459–83.

Dyck, I. (1996) 'Mother or worker? Women's support networks, local knowledge and informal child care strategies', in K. England (ed.) *Who Will Mind the Baby? Geographies of Child Care and Working Mothers*, London and New York: Routledge.

England, K. (1996) *Who Will Mind the Baby? Geographies of Child Care and Working Mothers*, New York and London: Routledge.

Flax, J. (1990) *Thinking Fragments: Psychoanalysis, Feminism, and Postmodernism in the Contemporary West*, Berkeley, CA: University of California Press.

—— (1993) *Disputed Subjects: Essays on Psychoanalysis, Politics, and Philosophy*, New York and London: Routledge.

Gergen, K.J., Gloger-Tippelt, G. and Berkowitz, P. (1990) 'The cultural construction of the developing child', in G.R. Semin and K.J. Gergen (eds) *Everyday Understanding: Social and Scientific Understanding*, London: Sage.

Grieshaber, S. (1998) 'Constructing the gendered infant', in N. Yelland, (ed.) *Gender in Early Childhood*, London and New York: Routledge.

Henriques, J., Hollway, W., Urwin, C., Venn, C. and Walkerdine, V. (1984) *Changing the Subject: Psychology, Social Regulation and Subjectivity*, London: Methuen.

Hewlett, S.A. and West, C. (1998) *The War Against Parents: What We Can Do For America's Beleaguered Moms and Dads*, Boston and New York: Houghton Mifflin.

Holloway, S.L. (1998a) 'Geographies of justice: pre-school-childcare provision and the conceptualisation of social justice', *Environment and Planning C* 16: 85–104.

—— (1998b) 'Local childcare cultures: moral geographies of mothering and the social organisation of pre-school children', *Gender, Place and Culture*, 5, 1: 29–53.

Hoogsteder, M., Maier, R. and Elbers, E. (1998) 'Adult–child interaction, joint problem solving and the structure of co-operation', in M. Woodhead, D. Faulkner and K. Littleton (eds) *Cultural Worlds of Early Childhood*, London: Routledge.

James, A., Jenks, C. and Prout, A. (1998) *Theorising Childhood*, New York: Teachers College Press.

Katz, C. (1991) 'Cable to cross a curse: the everyday practices of resistance and reproduction among youth in New York City', unpublished manuscript, Department of Environmental Psychology, City University of New York.

—— (1993) 'Growing girls/closing circles', in C. Katz and J. Monk (eds) *Full Circles: Geographies of Women over the Lifecourse*, London: Routledge.

Kirby, K. (1996) *Indifferent Boundaries: Spatial Concepts of Human Subjectivity*, New York: Guildford Press.

Kristeva, J. (1982) *Power of Horrors*, New York: Columbia University Press.

Maccoby, E.E. (1988) 'Gender as a social category', *Developmental Psychology* 24: 755–65.

—— (1990) 'Gender and relationships: a developmental account', *American Psychologist* 45: 513–20.

Oldman, D. (1994) 'Adult–child relations as class relations', in J. Qvortrup, M. Bardy, G. Sgritta and H. Wintersberger (eds) *Childhood Matters: Social Theory, Practice and Politics*, Aldershot: Avebury Press.

Robins, K. (1996) *Into the Image: Culture and Politics in the Field of Vision*, London: Routledge.

Rose, D. (1993) 'Local childcare strategies in Montréal, Québec', in C. Katz and J. Monk (eds) *Full Circles: Geographies of Women over the Lifecourse*, London: Routledge.

Sibley, D. (1995a) 'Families and domestic routines: constructing the boundaries of child-hood', in S. Pile and N. Thrift (eds) *Mapping the Subject: Geographies of Cultural Transformation*, London: Routledge.

—— (1995b) *Geographies of Exclusion*, London: Routledge.

Valentine, G. (1997a) '"My son's a bit dizzy." "My wife's a bit soft": gender, children, and the cultures of parenting', *Gender, Place & Culture: A Journal of Feminist Geography* 4, 1: 63–88.

—— (1997b) '"Oh yes I can," "Oh no you can't.": children and parents' understandings of kids' competence to negotiate public space safely', *Antipode* 29, 1: 65–89.

—— (1997c) 'A safe place to grow up? Parenting, perceptions of children's safety and the rural idyll', *Journal of Rural Studies* 13, 2: 137–48.

Winnicott, D.W. (1971) *Playing and Reality*, London: Tavistock.

Wood, D. and Beck, R. (1990) 'Do and don'ts: family rules, rooms and their relationships', *Children's Environments Quarterly* 7, 1: 2–14.

—— (1994) *The Home Rules*, Baltimore and London: Johns Hopkins Press.

Yelland, N. and Grieshaber, S. (1998) 'Blurring the edges', in N. Yelland (ed.) *Gender in Early Childhood*, New York and London: Routledge.

8

HOME AND MOVEMENT

Children constructing 'family time'

Pia Christensen, Allison James and Chris Jenks

Introduction

Social theory and contemporary cultural geographies have, over the last two decades, cast new light on social space. This has transformed our understanding of space from being either a neutral setting for social action, or a determined outcome of material conditions awaiting cartography, to a more deeply political conception of social space. Laclau (1979), for example, regards space as fixed and as antipathetic to change. The configurations of space that pertain in social relations are embodiments and manifestations of this stasis; such arrangements ensure stable cultural reproduction (Jenks 1993). What we 'know' as social space in our everyday lives, for example 'home', is thus a recognition of social order. As Harvey (1989) suggests, we are able to overcome the constraints of social space through conscious spatial arrangement. Space, then, is no opaque or inert medium, but must rather be understood in an intense and complex matrix together with identity, difference and differentiation. Social space is part of the process of identity-making and can, indeed, exhaust the possibility of identity. As Urry puts it, 'it is possible for localities to consume one's identity so that places become almost literally all-consuming places' (Urry 1995:2).

In this chapter we will engage with these ideas to explore how theories of home and locality and those focused on movement in contemporary societies may contribute to the broader understanding of the everyday lives and experiences of some younger children living in the North of England. We anticipate that the insights derived from our ethnographic study of children will have implications for theories of space, place and time. In this way we will demonstrate how research with children can contribute to social theory by interrogating its foundations through specific examples.

On the surface, home is a social space that appears dedicated to stasis and through popular imagery is constituted as such (Sibley 1995). Most material centring on social identity has seen home as the key source of rootedness. Such work explores the ways in which the concepts of self and personhood are acquired as particular ideas, modes and vocabularies of cultural belonging (Cohen 1982; Jenkins 1996). Traditionally, following the classic account of Barth (1969), the process of identity formation has been seen as tied to or reflected through particular fixed geographical or spatial localities. Within this tradition nationalist discourses, for example, have been regarded largely as debates about belonging, and social identity has been articulated through, and in relation to, tangible, material spaces. Cohen (1982) suggests that this emphasis on firm boundaries has led to a persistence, and almost a reification, of the idea of the group or of ethnicity. Jenkins similarly criticises this tradition for 'espousing a middle-of-the-road materialist realism that resonates with the core themes of pragmatism' (Jenkins 1996: 99) and for working with a determined causality that stems from an assumption of the individual's influence upon the collectivity.

In opposition to the Barth thesis of 'home' as a placed identity locus there is, however, a more contemporary literature concerning human movement. This burgeoning corpus critiques the fixing of identity in relation to particular spaces or localities, or indeed in relation to particular times or historical moments. This is the vehicle of the decentred identity of post-structuralism and the post-historical being of postmodernism. Hebdige prefigured these arguments well:

> a growing scepticism concerning older explanatory models based in history has led to a renewed interest in the relatively neglected, 'under-theorised' dimension of space . . . It has become less and less common in social and cultural theory for space to be represented as neutral, continuous, transparent or for critics to oppose 'dead . . . fixed . . . undialectical . . . immobile' space against the 'richness, fecundity, life, dialectics of time', conceived as the privileged medium for the transmission of the messages of history. Instead spatial relations are seen to be no less complex and contradictory than historical processes, and space itself refigured as inhabited and heterogeneous, as a moving cluster of points of intersection for manifold axes of power which cannot be reduced to a unified plane or organized into a single narrative.
>
> (Hebdige 1990: vi–vii)

Within such postmodern discourse it is argued that, rather than fixity, 'movement has become fundamental to modern identity and an experience of non-place (beyond a "territory" and a "society") an essential component of everyday existence' (Rapport and Dawson 1998: 6). This feature, it is suggested, has an overall importance for appreciating the complexity of modern experiences and practices in the late twentieth century. Within anthropology, for example, the emergence of this perspective has entailed a critical examination of the discipline, bringing its epistemological grounds and elementary practices under scrutiny. It has been argued that the traditional anthropological interest in identifying and localising groups and societies of socio-cultural significance underplays the complex relations of, and between, time and space. Anthropological studies have tended to elicit and represent cultural separateness rather than cultural and societal interconnectedness, and to see social relations and cultural practices as fixities rather than unfinished, in progress or made *en route* so to speak.

These new perspectives have been developed particularly in relation to studies of migrants, refugees, travellers and diasporic communities, and thus have become centrally engaged in the study of the 'making of home away from home' (Clifford 1997; Rapport and Dawson 1998). Building on the notion of space as a social construct and thus home as primarily an analytical construct through which individual identity can be explored – for 'home is where one best knows oneself' – Rapport and Dawson argue strongly against static conceptualisations of identity formation in relation to fixed spaces or frozen times:

> the emphasis on a relationship between identity and fixity has at least been challenged in anthropology of late by representations of the relationship between identity and movement. Now we have 'creolizing' and 'compressing' cultures and 'hybridizing' identities in a 'synchronizing' global society. Part of this reconceptualization pertains significantly to notions of home: part-and-parcel of this conceptual shift is a recognition that not only can one be at home in movement, but that movement can be one's very home.
>
> (Rapport and Dawson 1998: 27)

Rapport and Dawson thus argue that traditional conceptions of individuals as members of separate localised communities and insulated cultures have become increasingly redundant in a world characterised by the accelerating processes of globalisation, synchronicity and hybridisation. The call is for social theorists to engage with the practices and configurations of identity in and through relations of time and space and, as they suggest, to regard the search for the modern self 'as inextricably tied to fluidity or movement across time and space' (Rapport and Dawson 1998: 4).

This emerging emphasis on the dynamic and changing temporal qualities of identity formation has permitted the recognition that identities are neither fixed nor given; they are not ascribed by belonging to a particular culture or living in a particular space. Rather, identities are achieved, negotiated, experimented with and challenged. However, the extent to which such a shifting and temporalising construct must, necessarily, lead to an abandonment of *any* or *all* spatial

141

markers of identity, we would want to question in line with the views of Keith and Pile (1993):

> it may be argued that simultaneously present in any landscape are multiple enunciations of distinct forms of space – and these may be reconnected to the process of re-visioning and remembering the spatialities of counter-hegemonic cultural practices. We may now use the term 'spatiality' to capture the ways in which the social and spatial are inextricably realized one in the other; to conjure up the circumstances in which society and space are simultaneously realized by thinking, feeling, doing individuals and also to conjure up the many different conditions in which such realizations are experienced by thinking, feeling, doing subjects.
>
> (Keith and Pile 1993: 6)

The child at home in the family

Dawson and Rapport (1998) rightly note that the concept of 'home' is a conceptual construct which guides our actions and informs our identities. It must, however, be remembered that it can be nonetheless (and indeed often is) a physical, spatial context. These twin aspects of home, as James (1998) has argued, have been employed historically as cognitive resources through which, over time, children have been made to be literally 'at home' in the family. One of the legacies of the industrial revolution was, for example, the separation of the work place from the home, leading to the gradual recognition of the 'home' as the key physical and emotional setting for people's personal and private lives. In turn, this 'home' came to contextualise familial experiences, what Allan and Crow (1989: 1) term 'the modern domestic ideal' of parents and children living within the nuclear family. The gradual centring of 'the child' within 'the family' meant that by the late 1800s the family was no longer regarded simply as the site for reproduction, but had taken on responsibility for socialising (civilising) 'the child':

> the concept of childhood dependency . . . developed as a feature of the merging nuclear family and was justified through recourse to a cultural mythology which was developing simultaneously in relation to the child . . . Children's social dependency was fast becoming a key feature both of the family and of childhood itself.
>
> (Hockey and James 1993)

And in turn, 'the child' became increasingly dependent upon its family and its home for, as Oakley notes, the two terms are used as increasingly synonymous: 'the home is the family' (cited in Allan and Crow 1989: 2).

Contemporary family life in Europe and North America is constituted through the union in marriage or partnership between two ideally independent persons who on becoming parents, are seen as having a determining role in the creation of the child as an individual. In the course of growing up the child is expected to develop as an individual through achieving its own independence (Christensen 1994). An imperative to this maturation of the individual self is the child's separation from his or her parents, who are equally seen as responsible for this latter process to take place. The symbolic marker concluding this separation process is irrevocably linked to the physical space of the home: when the child in youth moves from their family home to his or her own first accommodation, the basis for creation of his or her own family, then that child is regarded as competent, mature and independent. Thus, should the parental home still carry out or provide different household support for the child, it is commonly perceived as prolonging the dependency of the child. Through the familial social bonds of care and love, the physical space of the house thus becomes transformed in and through time into a 'home', the space and place where identities are worked on: children develop their social competence and demonstrate and enact their growing maturity at home and in movements in and away from this space.

In the same vein, Mary Douglas (1991) has insisted on the importance of time in the making of homes, arguing that homes structure time and memory through their capacity to spatially order and control the activities of family members via such things as communal eating, the division of labour, moral obligations and the distribution of resources. For her, 'the family' and 'home' are both sites within which notions of togetherness and community (including a sense of belonging) are articulated and established under different material, ideological and emotional conditions. In this chapter we draw on these ideas to argue that the local domestic space of the 'home' is a key site for children's experience and understanding of 'the family', realised in and through everyday negotiations about time and space. In particular, however, we will argue that children's understanding of themselves and of their family is achieved through the movement in, out and around the home of different family members as much as it is through the 'home' as a material space and a fixed locality. Through prosaic and intricate contestations over the use of physical space of the home, children are enabled, as individuals and over time, to understand what family time, and their own contribution to it, means.

The study

This chapter draws on data from an ethnographic study of children's understanding and use of time. The study focused on 10–12-year-olds and their experiences of the transition from primary school to secondary school and took place in three primary schools and two secondary schools in an urban and a rural setting in the North of England. We discuss the data we gathered about children's experience and use of time at home with parents, sisters and brothers through conversations with children at primary school. The primary schools included a small village school that had 57 pupils, two other primary schools – one in the city centre of a large town and another in a bigger village. These schools were of comparable size, each with about 225 pupils. A wide range of methods was used to capture children's conceptions and practices of everyday time at home and at school; social relations at home with regard to time use and the negotiations this involves; time values and social organisation; biographical time including issues of age and development. The methods used included individual and peer group interviews together with ethnographic participant observation in schools and, later, a large-scale quantitative survey.

However, in addition to these more conventional research techniques, a set of participatory research tools was developed to explore how children come to embody the temporal practices and rhythms through which their everyday lives unfold (Christensen and James 2000). Data was produced using a set of simple paper charts, inscribed with circles, boxes or lines which the children filled in as they chose. These charts provided a way for us as researchers and the children to concretise what are often rather abstract or implicit ideas about time. For example, through creating their own pie charts to depict the way time is spent in the family, with friends or on one's own, the children were able to reflect on both the pictorial representation and, in discussions on the experiential differences between different kinds of times, their own control over time use and their social relationships. Thus what is usually experienced as taken-for-granted everyday social practice was, through the use of the tools, granted an unusual and explicit solidity and its familiarity and routine were temporarily objectified and estranged.

Family time at home

In the children's accounts, the concepts of family time and family life were articulated through two perspectives on 'home' which confirm the settledness of 'family' with 'home'. First, the 'house' was regarded as a physical location, a physical space where children centred themselves. In describing their everyday lives, the children would talk of 'my house' or 'his house', thereby firmly locating themselves and their peers in physical as well as social space. Moreover, on their charts the children clearly distinguished between activities which take place inside and outside the house and, in conversation, they would constantly refer to the to-ing and fro-ing which takes place between the children's houses. Keith describes this pattern of activities:

Like Tim comes to my house straight after school and Tim's mum
brings Tim's youth club clothes round to my house and then we both
get changed and we play on the computer and stuff. And then he stays
for tea at my house on a Friday and then we both go to youth club
together and then my mum picks us up.

Second, and reaffirming this spatiality of the house, children recognised that
the 'house' symbolises and gives practical and material voice to the idea of 'home'
as the locus for the social relations which comprise the family. This is highlighted by the children's accounts of their own and other family members'
movements in and out of both the 'house' and the 'family' in the course of their
everyday lives.

A starting point for our analysis, therefore, is to challenge the suggestion that
the movement and fragmentation which characterise contemporary family life is
destructive for children as is publicly represented. On the contrary, as we will
argue, it is part and parcel of children's everyday lives and formative of their
social learning. Qvortrup (1995), for example, has argued that the experience
of fragmentation signifies everyday family life in contemporary society and sees
this as a growing generational problem. Changing employment patterns, family
breakdowns and an increased institutionalisation of children's everyday lives
(cf. Aitken, Chapter 7) has brought about a decrease in the quantity of time
spent together by families, leading in turn to an emphasis on the importance of
the quality of the time that children and parents do actually spend together.
Today, it is argued, parents and children spend more time apart than the previous
generations did and therefore, because everyday family time together is scarce,
spending 'quality time' must substitute for this lack. In Denmark, for example,
this notion had particular prominence in the early 1990s. These points were not
only widely discussed, but a national campaign, run by staff in day-care institutions and their union, promoted the slogan 'Have you talked with your child
today?' This reminder greeted parents on posters on the entrance-door to the
day care institution and was also displayed on walls and on car stickers. This
was in line with previous research which showed that parents readily define
'togetherness' in and through time (Christensen 1999). Thus it could well be
argued, as Qvortrup implies, that changes in the amount of time that children
and parents spend together will necessarily influence the symbolic meaning of
that 'togetherness'.

However, though our data on the surface would seem to confirm and reflect
this contemporary familial ideology – it revealed, for example, that an over-
whelming majority of children liked to spend time with their family at home,
best of all – at the same time, and most importantly, it begged the question of
what exactly 'quality time' means for children. For children, we will show, time

with their families represents something different from the notion of 'quality time' expounded by parents, educationalists, researchers and policy makers. As our data shows, children do not attempt to quantify family time; rather it is experienced largely as an undifferentiated and unremarked temporal flow which is primarily realised through the social spaces of the home.

Some children, for example, found it very difficult to articulate in conversation with us how exactly they spend time with their family. They would struggle to tell us what they actually did together. When pressed, they might identify digging the garden, cooking, watching TV or visiting their grandmother. For the most part, however, in their accounts, family time at home simply *is*. The difficulty of putting into words what this time might consist of appears in the following statement given by Leanne: 'Sometimes we go out or sometimes we just stay in.' Her statement captures well the very ordinariness and non-eventful quality of time at home in the family, which the other children expressed in similar terms. Their evaluation stands in marked contrast to the usual adult formulation of 'quality time', envisaged as a very clear set of activities and experiences, designed to demonstrate forms of loving behaviour, togetherness and care, all of which are understood to be ways to strengthen the family bonds which are threatened by the continual movement of family members in and out of the house and in and out of each other's lives.

What, then, is quality time for children? Does it necessarily involve the interconnection of time and space? For some children, a sense of togetherness and belonging is established in everyday life by family members simply being together in the same space. One girl, Julia for example, reflecting and puzzling on what this togetherness might mean, eventually described it as follows: 'Sometimes I do stuff *with my family* . . . we don't really do anything *as a family*.' Here the distinction drawn between 'doing with' and 'doing as' is important. It emphasises togetherness as the sharing of space through doing something with one's family. The more self-conscious enactment of the idea of the family – doing things as a family – is less important in her account. This theme was also illustrated by Thomas who, in talking about how he spent his time, did not think it appropriate to identify a specific portion of his week as time spent with the family. His explanation for the omission of 'family time' from his pie chart was revealed in our later conversation.

THOMAS: Family is something that's *there* – every day – you're sort of used to them and then it becomes normal (ordinary) and you don't really think about it.

INTERVIEWER: But, what about the school, that's also every day!

THOMAS: Yes, but I have also written that in.

INTERVIEWER: Yes but how come then the family?

THOMAS: Oh, but the school that's different *because that's in another place than right where I am.* I mean my family almost every time of the day they are with me. I mean one of them my mum, my dad or my little sister.

Such data, drawing as it does on spatial concepts, reveals the extent to which the home (as a concept as well as a physical space) is integral to children's notions of family time and is as much, or even more, enacted in everyday comings and goings as it is constituted through special events or times.

This is not to suggest that children did not identify particular times when families are together and sharing the same space, but these were rarely picked out for their special or non-ordinary qualities in the flow of everyday life. For example, in some families the evening meal had particular symbolic meaning through being the only time the family actually was together in the same place during hectic weekdays. The communality of the meal created for the family a symbolic transition between work and home and between the personal world and the community of the family. The family was brought together after having been separated in each of their individual activities. After the meal the togetherness was (temporarily) dissolved, for family members would again engage in their individual activities. Describing how they spend their time after school, most children would depict the evening at home as a set of movements in and out of the home, shaped by the temporal structuring of meal times. John explained: 'I come in about half past five for my tea and then I go back out again.' For the children, tea-time is family time. It is the time when contact with peers must be interrupted; they had to come in from the street to eat and be together with their families.

Watching television offered a comparable instance of the creation of family time in the space of the home. Many children, for example, identified as significant the togetherness of sharing the sofa with their parents and siblings while watching TV and as something which represented time with their families. It was something which they often did in spite of the fact that many of the children had their own TVs in their bedrooms. But this was not necessarily a time of familial harmony. This togetherness often involved disputes and confrontations around control over the space and material objects through which 'home' is constituted, as Martin described:

and when we're watching telly at my grandmas, if I was watching summat and I turn it over and my sister was watching what I turned over, she wanna watch the thing that was on before, you know. I get told off . . . if she was watching summat she goes and gets a drink or summat, I turn the telly over and she screams and has a fit. I just shout me Mam and me Mam tells Tina to go and watch her own telly.

147

These routine disputes pattern family life and were not seen by children as special or remarkable: they are simply what happens at home in the family, routine events among other routines, as Rebecca described: 'So, every Sunday everybody goes down to my Auntie Tracy's house and my mum and Auntie Tracy *always* cook all the tea.'

The temporal patterning of the comings and goings of family members in and out of the home and around the physical space of the house represents, we argue, the entry of children into and out of family time. Through such movement – for example, being 'allowed' out of home, 'having to' come in for tea – children negotiate time on their own and achieve a growing sense of independence from the family and, at the same time, help constitute the notion of family time. Thus, while clearly revealing the importance of movement for an understanding of the meaning of home, this ethnographic case also, necessarily, requires for its explanation a conceptualisation of the *spatial domain* of the home. Thus for example, the spaces within the home – the kitchen, the dining table, the bedroom, the sitting room, the sofa – are locations where children come to realise the boundaries between own time and space and family time and space (cf. Sibley 1995). This will be discussed in the final part of this chapter.

These disputes between family members over time and space in the home, we suggest, are inextricably linked to the tension underlying the values attached to the rights, privileges and independence of family members and those which promote family 'togetherness', for example reciprocity, mutual responsibility and familial solidarity (Christensen 1999). Families are only able to manage to contain such contradictions, we argue, through a constant balancing of the concerns for independence and that of togetherness, a process articulated in and around particular spaces in the home. The welfarist call for parents to spend *more* 'quality time' together with their children is therefore potentially problematic if family 'quality time' denies children their need to be both with and *without* their families, as we discuss in the next section.

Own time

Solberg (1990), in a study conducted among Norwegian children, showed that the time children spend while being alone in the home after school, and before the rest of the family arrived home, had importance not only for their perception of having control over their own time, but also contributed to their general experience of independence. But the aloneness to which she refers was that of the absent parent. Children were able to spend time in each other's homes, with their friends, without the watchful parental gaze. This they valued.

However, our study showed that the concept of own time or time on one's own is ambiguous for children and, indeed, many children revealed that they did not like to spend much time on their own at home. This ambiguity partly stemmed from their interpretation of what own time means. For some children the question 'How much time do you spend on your own?' signified their exclusion from the social world of their peers, and others more generally, rather than representing, as in Solberg's (1990) case, the freedom to make an independent choice over time use. Georgina, for example, commented as follows: 'You've got no one to talk to.' Tim agreed: 'It's boring. I don't have nowt to do.'

Indeed, when piloting our research tools, we were advised by one boy to recast a question asking about time spent alone and instead to ask about time spent on 'my own'. He said that the former would indicate that a child was a 'sad' person; to be alone signifies someone who does not have any friends, he explained. In this subtle distinction we see an important idea: time alone is, for children, not voluntarily chosen, but imposed and defined by social relations taking place in time *and* space. This interpretation was confirmed by Janice. She said that she hated being on her own because that meant she was alone and she didn't want to be seen as a 'Billy No Mates'.

In the children's accounts, sometimes time on one's own was relished and deliberately chosen. Such control over time, and making it one's own, is something which children learn to do, and in this process the spatial dimensions of the home provide a key site. The home, for example, provides children with a context within which to begin to exercise such control over time use for themselves. Children were able to create their 'own time' out of 'family time' through moving in and out of particular spaces in the house. In their accounts the children described, for example, how they hid in their bedrooms to avoid participating in domestic tasks or doing homework. One boy, William, explained: 'I go upstairs, I try and run away as well.' It was as if they acted on the general saying 'out of sight, out of mind', and from the children's point of view, thus possibly were able to escape the control of their parents. Christopher, telling the researcher about doing his homework, said: 'I go upstairs and I intend to do it (homework) and get in bed and when I hear someone coming upstairs I just jump out of bed and do it.' The relative privacy of the bedroom makes family time yield momentarily to own time and provides a space for children to resist what they saw as the bossiness of their mothers and what they experienced as the more generalised displacement of their own agendas by those of their parents. Keith described how he and his brother managed to use their computer 'after hours': 'My little brother turns the sound down, so mum can't hear it.'

What does having time on one's own, mean to children? For some, as in the examples already given, it signifies a growing sense of personal independence from the family, an independence made possible through children's literal movement from one space within the house to another. To remove oneself from the social space of the family was a strategy used by children to find a space within which, on their own, they could express and cope with bad moods or emotional tensions. In this sense, therefore, such movement in space could be said to image children's awareness that the expression of strong negative emotions must be temporally and spatially constrained. They recognised that the visible and explicit show of emotions in public is not culturally permissible or appropriate. Two girls in a conversation said:

CLARA: The only thing (time) when I'm on my own is when I'm in a mood.
KERRY: I only go on my own when I get told off and then I just cry cos I feel sorry for myself.

What is apparent in these examples are the processes by which children learn to manage the tensions between 'self' exclusion and the 'social' exclusion which being 'alone' represents.

This was further confirmed by parents who insisted that children should participate in 'family time'. That children chose to be on their own by removing themselves from the social space of the family was seen as problematic and often contested by parents, as can be seen in the following example. Eleanour said:

It's like my brother and he gets a real mood and he goes off in a mood. So me Mam tells him to get back downstairs and watch telly. And then we just start arguing and he calls me a poof and I call him a poof and we just start hitting each other, he punches me in my arms so I punch him back.

However, alternatively, parents may make children 'be on their own' to punish them. The concept of 'grounding' was discussed often by the children. 'Grounding' forcibly removes children from the social world of their peers, ironically by symbolically 'placing' them on their own, but centrally within the space of the family home. In this way, 'time on one's own' is used by parents as a form of social rather than self-exclusion. This adds to the ambiguity with which children regard 'time on one's own'.

As noted earlier Douglas (1991) drew attention to the importance of the physical space of the home. She also highlighted the way in which the essential routines and regimentation of home life and home-making may ironically deny the individual freedom of choice or self-expression through the inevitable, and in many cases purely pragmatic, exertion of power by some family members over others. Douglas also drew attention to the fact that the routinisation of household practices produces a sense of solidarity. What is now called 'quality time' can be reduced in definition to enforced proximity, dedicated interaction, and the exclusion of non-family members periodically within the geography of private space. In our data we can see these processes taking place, and they are processes in which children themselves play a major role. Childhood agency manifests itself then, in the negotiation of the texture and periodicity of quality time.

Quality time as family time in family space

Traditionally, the power to designate routines and to enact control over household schedules and practices would be taken to be a systematic and unproblematic feature of parental care of and control over children. Indeed, we would point to the fact that the British government has recently introduced parenting classes to ensure that this convention is continuously re-enacted so as to ensure or reinvigorate the idea of 'the family' unit. However, the children in our study, and by implication in wider society, locate themselves firmly as proactive within this discourse of power. Children project themselves as actors with a dynamic role to play in contestations over time and space. A main arena for this is the dispute between 'family time' (ideologically expressed as 'quality') and 'own time'. Such disputes are played out through the parameters of age and mediated through strategies of resistance within the family.

Parents, for example, may exert the moral imperative that time is money, that time on one's own should be spent actively and productively. Marcia said:

> Me Mam, she always tells me to move or to go and get something to do, while I am sitting there watching the telly. But while I'm by myself, I don't have to. I can just sit there watching the telly or go on the computer.

Kit revealed a similar experience:

> She'll say 'go outside and lark with your mates' or something like that. Or she'll say 'watch telly for ten more minutes and then go out'. But I don't.

In these examples, own time is not yet something which parents consider their children to be able to be responsible for. They see their 10-year-old children as still needing some supervision over how they spend their time at home. Children, therefore, have to seize those opportunities for shaping their own time at home when they arise. Sally told the researcher:

> Me Mam sometimes go out, she goes to the shop and leaves me on my own . . . she doesn't shout at me . . . you can do whatever you want and there's no one to tell you not to do that.

151

Kenny agreed: 'It's quiet and you get to do what you want to do.'

That some children at this age are beginning to feel the need for time on their own at home, as a contrast to their busy everyday lives at school and a rest from sports activities, can be seen in Paul's vivid portrayal:

> You are made to make do decisions, it's like, cos when you're at home you can lie back, you can do anything. But when you're at school it's, you've got to do maths, you've got to do English, you've got to do that, you've got to do this. I don't really like doing that, I think it's a bit boring but I have to.

Steve's account resonates with this. He said: 'You really just want to keep one (day) to yourself.' Charlotte echoed both boys:

> Just non-stop working and you're just going like this (sighs) and just writing everyday and it just gets real boring and sometimes you think: God, can't we have a week off or something? cos it just gets real tiring . . . sometimes when it's boring, it just feels as though the day never gonna end.

Another boy, Tim, already recognised how the patterning of time changes over the life course and gave us an account where his notion of time resonates with that of Frykman and Lofgrens (1979): that time has become an object, working almost as a force with a life of its own that can dominate and restrict people's lives.

> I know summat, when you get older, you lose, time to do things you wanna do, don't you. Yeah, gets busier and you can't do anything you wanna do.

From these examples it is clear that age itself is an important element in the negotiations which take place around the spatial and temporal ordering of the home and the implicit age hierarchy that it embodies. It means that children's wishes and feelings about time use become subject to the whims of parental wishes and authority, the structure of the family and the work patterns of parents. Lilly said:

> My mum woke me up at half past six today to do my bedroom. Yeah it was a dump. I had the cupboards all open, I had like me best clothes all over the floor.

Max amplified these felt constraints:

152

cos my mum has to go to work and my dad doesn't live with us, I have to get up at half, quarter past seven at the latest. I never do to get . . . to go to my child minders' but on the weekends or holidays I can just play on the computer until 9.00, get, spend an hour getting myself ready like, choosing what clothes I like.

But children's claims over own time and space were also the subject of ongoing fights and quarrels with siblings about the use of television, getting time on the computer, general untidiness or who is going to tidy up one's bedroom, help with setting the table and washing up. Elena, for example, described in detail the complexity of her relationship with her 14-year-old brother and their managing of the maternal gaze when they have their fights at home.

Sometimes we, I, usually have an argument and I usually come and, erm. I'm naughty and stuff and [my] brother goes 'I'm sick of you, you're the one who gets me into trouble all the time'. Then I go upstairs and beat him up and he goes 'Ahh, Elena stop it'. He does, honest, he does and, erm, he goes 'Elena will you stop doing that, you're messing it all up'. And I say 'Excuse me, but look inside your bedroom. It's a tip!!' And he goes 'It's not!!' and it's full, of course. My mum comes upstairs 'Right, what are you two up to?' My bed's a mess because we've [been] punching him and I tidy me hair up and then I make my bed and I'm, I go 'Nothing mum'. Then she goes in Robin's room and says 'Robin you are a scruffy pig'. He's got this poster thing and it says: My Room, My Mess, My Business.

Conclusion

What then does this ethnographic account contribute to social theories of home? First, it restates the importance of material space for concepts of belonging and identity. This does not, however, mean that a static notion of home or a fixed identity thereby follows. Rather, as we have shown here, it is precisely through the dynamic and fluid movement of children in, out and around the home that their own sense of belonging to the family and the home is constituted. This also points back to the political, and in many senses, intentional character of social space as considered in the introduction to this chapter. Identities are forged within a locational matrix of constraint, contested meaning, conventions of placing and avenues of possibilities. As we stated, spatial relations and the social relations they contain are essentially dynamic.

Second, and related to this, is that through this movement 'family' is itself constituted – that is to say, 'the family' as an idea, is continually made and remade by children and parents through the negotiation and juxtaposition of time in space. This shows that control and power are fundamentally exercised by the movements and positionings through which emergent identities are made and articulated in people's everyday lives. This point is made forcefully by Soja and Hooper:

> The cultural politics of difference, whether old or new, arise primarily from the workings of power – in society and on space in both their material and imagined forms. Hegemonic power does not simply manipulate naively given differences between individuals and social groups, it actively *produces and reproduces difference* as a key strategy to create and maintain modes of social and spatial division that are advantageous to its continued empowerment. At the same time, those subjected, dominated, or exploited by the workings of hegemonic power and mobilized to resist by their putative positioning, their assigned 'otherness', struggle against differentiation and division. This socio-spatial differentiation and struggle is, in turn, cumulatively concretized and conceptualized historically and geographically as *uneven development*, a term which we use to describe the composite and dynamic socio-temporal patterning of socially constructed differences at many different scales from the local to the global.
>
> (Soja and Hooper 1993: 184–5)

Third, this study reveals the importance of time itself for ideas of belonging and home, most obviously here represented through age/maturity as conditioning access to space, a contestation which takes place *in* and *over* time. In this sense 'home' and ' belonging' are temporally as well as spatially enacted. We would argue, therefore, that movement and change are characteristic of children's everyday life. Thus the concern to attach children to the home, which arises through a further concern about the fragmentation of children's lives, is questioned by our study.

Acknowledgement

This study was funded by the ESRC under the Children: 5–16 Research Programme.

References

Allan, G. and Crow, G. (eds) (1989) *Home and Family: Creating the Domestic Space*, Basingstoke: Macmillan.

Barth, R. (1969) *Ethnic Groups and Boundaries: The Social Organisation of Culture Difference*, Oslo: Universitetsforlaget.

Christensen, P. (1994) 'The child as cultural other', *KEA: Zeitschrift fur Kulturwissenschaften, Tema: Kinderweltern* 6: 16.

—— (1999) 'Towards an anthropology of childhood sickness: an ethnographic study of Danish schoolchildren', unpublished Ph.D. thesis, Hull University, UK.

Christensen, P. and James, A. (2000) 'Childhood diversity and commonality: some methodological insights', in P. Christsensen and A. James (eds) *Research with Children*, London: Falmer Press.

Clifford, J. (1997) *Routes: Travel and Translation in the late Twentieth Century*, Cambridge, MA: Harvard University Press.

Cohen, A.P. (1982) *The Symbolic Construction of Community*, London: Routledge.

Frykman, J. and Lofgren, O. (1979) *Den kultiverade manniskan*, Lund: Libers Forlag.

Douglas, M. (1991) 'The idea of home: a kind of space', *Social Research* 58, 1: 287–307.

Harvey, D. (1989) *The Condition of Postmodernity*, Oxford: Blackwell.

Hebdige, D. (1990) Introduction to special edition 'Subjects in Space', *New Formations* 11: vi–vii.

Hockey, J. and James, A. (1993) *Growing Up and Growing Old*, London: Sage.

James, A. (1998) 'Imaging children "at home", "in the family" and "at school": movement between the spatial and temporal markers of childhood identity in Britain', in N. Rapport and A. Dawson (eds) *Migrants of Identity*, Oxford: Berg.

Jenkins, R. (1996) *Social Identity*, London: Routledge.

Jenks, C. (1993) *Cultural Reproduction*, London: Routledge.

Keith, M. and Pile, S. (eds) (1993) *Place and Politics of Identity*, London: Routledge

Laclau, E. (1979) *Politics and Ideology in Marxist Theory*, London: Verso.

Qvortrup, J. (1995) 'Childhood and modern society: a paradoxical relationship', in J. Brannen and M. O'Brien (eds) *Childhood and Parenthood*, London: Institute of Education.

Rapport, N. and Dawson, A. (1998) (eds) *Migrants of Identity*, Oxford: Berg.

Sibley, D. (1995) 'Families and domestic routines: constructing the boundaries of childhood', in S. Pile and N. Thrift (eds) *Mapping the Subject*, London: Routledge.

Soja, E. and Hooper, B. (1993) 'The spaces that difference makes: some notes on the geographical margins of the new cultural politics', in M. Keith and S. Pile (eds) *Place and the Politics of Identity*, London: Routledge.

Solberg, A. (1990) 'Negotiating childhood: changing constructions of Norwegian childhood', in A. James and A. Prout (eds*)* *Constructing and Reconstructing Childhood*, Basingstoke: Falmer Press.

Urry, J. (1995) *Consuming Places*, London: Routledge.

9

TRANSFORMING CYBERSPACE

Children's interventions in the new public sphere

Gill Valentine, Sarah L. Holloway and Nick Bingham

Introduction

Information technology is one of the fasting growing sectors of the economy. Over 40 per cent of US households now own a home personal computer (PC) (Vanderkay and Blumenthal) while the recent UK Government's IT for All survey (UK Government) found that 34 per cent of UK respondents claimed to have a PC at home even if they did not use it. Families with children have the highest levels of ownership. Although the domestic PC market which started in the 1970s was initially the domain of hobbyists, in the early 1980s US and UK government initiatives to put computers into schools inspired the computer companies to target the consumer market (Murdock, Hartman and Gray 1992). While computer games, entertainment packages and the opportunity to surf the Internet hooked children's interest, companies also picked up on government discourses about the impending 'Information Age'. Information and communication technologies (ICT) have been marketed as 'frontier technology' with prompts that the sooner your children develop technological skills, the better their future educational and employment prospects will be (Haddon 1992). As Nixon explains:

> the family is being constructed as an important entry point for the development of new computer-related literacies and social practices in young people . . . what is discursively produced within the global cultural economy as digital *fun and games* for young people, is simultaneously constructed as *serious business* for parents.

> (Nixon 1998: 23)

156

Yet, at the same time as parents are being urged to develop their children's technological competence, this very competence at using ICT is alleged to be putting children at potential risk. The relatively unregulated nature of cyberspace means that soft and hardcore porn, Neo Nazi groups, paedophiles, racial and ethnic hatred can all be found on the Net (Kitchin 1998; Squire 1996), thus prompting fears that children may be corrupted by the unsuitable materials that they can find on-line or meet dangerous strangers in cyberspace. As Wilkinson (1995: 21) points out, 'we are all having to face up to the fact that our children's familiarity with technology is bringing a new set of risks, especially if we want them to take full advantage of computers as tools of empowerment and education'.

In this way, discourses on children as both technologically competent and at risk from their technological skills, are being used to mobilise other understandings of childhood more commonly associated with public space, namely that young people are 'vulnerable', 'innocent' and in need of protection from the adult world (Valentine 1996). At the same time, by constructing children as deliberately seeking unsuitable materials on-line, these discourses also reproduce another common representation of childhood evident in debates about public outdoor space – that of the 'dangerous child' (Oswell 1998). This imagining of childhood represents children as troublesome, innately sinful and in need of discipline and control (Valentine 1996).

The relationship between children and technology is also a key focus of concern in current popular debates about children's use of space and time. Some commentators argue that computers are a sedentary activity and that children are spending too much time indoors using ICT (through choice and because their parents are using these attractions to structure their use of time in order to prevent them playing in public space unsupervised because of traffic and stranger-dangers), instead of getting physical exercise playing outdoors in public space. As a consequence, technology is being blamed for undermining children's social relationships within the household and their friendships within the local community, as well as robbing them of the capacity to enjoy the sort of imaginative outdoor play which adults recall from their own childhoods.

All these concerns are undercut by a more fundamental anxiety: that children's technological skills may outstrip those of their parents. As a consequence, some adults fear that if they are less technologically competent than their offspring, then not only may they lose their ability to control their children's computing activities, but their 'natural' authority or power base in the household will also be eroded.

Yet, there are two fundamental problems with the way these popular fears are being peddled in the media. First, they are technologically determinist in that they assume that technology impacts on children's lives in negative or harmful ways. In other words, particular outcomes are being attributed to ICT which ignore the way that the impact of the technology may vary according to specificities of time and place, who is using it, their intentions and the other agendas to which technology may become attached (Bromley 1997). It is what Bryson and de Castells (1994: 207) term an 'artifactual view' where technology is severed from the normative context of social practice.

Second, these fears pay no attention to children's actual uses or experiences of ICT. While adult 'folk devils' can often resist the ways that they are constructed in moral panics (McRobbie and Thornton 1995), for example by speaking out in the press, children rarely have the same power or opportunities to challenge representations of their use of the Internet in the adult-controlled spaces of the media. As Waksler (1986) has argued, the contemporary Western construction of childhood is one in which children are commonly produced as less developed, less able and less competent than adults. In this way children's agency and their sophistication at negotiating and managing their own lifeworlds are often overlooked (Alanen, 1990; Prout and James 1990).

This chapter gives children a voice in these debates by exploring their own approach to potential on-line risks and the extent to which ICT shapes their use of domestic space, time and social relationships. In doing so we start from the assumption that there are no set outcomes produced by some internal technological logic, rather that the properties of ICT emerge in practice as they are domesticated by individuals and households in what is a reciprocal process of change (Silverstone, Hirsch and Morley 1992).

The chapter is based on material collected as part of a two-year ESRC-funded study of children's use of ICT at school and home. The first stage of the research was based in three secondary schools in the UK (the names of the schools and interviewees have been changed to protect their anonymity). Two of the schools are in a major urban area in South Yorkshire, the other in a rural coastal town in Cornwall. The two urban schools were divided in terms of their intake: whereas Highfields draws mainly on a 'middle-class' catchment area, the parents of children at Station Road are primarily 'working-class'. The Cornish school, Westport, has the socially mixed intake which is often found in rural areas. Within the case-study schools we undertook a questionnaire survey of 753 children aged 11–16 asking about their use of computers and the Internet in both school and home environments. This was followed by observation work in a number of case-study classes and focus group discussions – based mainly on existing friendship groups – which covered children's experiences of IT within the school environment. Semi-structured interviews with the IT- and head-teachers from these schools were also carried out.

On the basis of this work in schools and electronic snowballing to locate high-end users, forty children and their families were asked to participate in a further stage of the research. This involved separate in-depth interviews with the parent(s) and the children in the household about the purchase of home PCs and Internet connection, use of computers and the Internet by different household members, different competence levels, issues of unity or conflict around shared use, owner-ship, the domestic location of the PC and control of its use and whether being on-line had affected household relations. It is this second stage of the research upon which this chapter draws.

Cyberporn!

In 1995, the US magazine *Time* rang a cover story titled 'Cyberporn!' (3 July 1995) in which it claimed that 83.5 per cent of the images on newsgroups were sexually explicit. The same year a German federal prosecutor cautioned CompuServe, a major Internet service provider, that over 200 Internet news-groups accessible through CompuServe allegedly broke German law (Cate 1996; Markoff 1995). Such reports represent only the tip of the iceberg. There has been an avalanche of studies and newspaper reports about the extent of sexu-ally explicit material available on the Internet. These reports commonly represent on-line and off-line space as two distinct worlds. The off-line world – particu-larly the home – is imagined as a space of childhood innocence where children are assumed to have no access to pornography or unsuitable materials. Instead, these materials are imagined to be contained within an 'on-line' world which contaminates the so-called 'real world' by breaching the sanctuary of the home, invading and polluting it with sexually explicit images and 'dangerous information'.

It is an imagining of both 'childhood' and on-line and off-line space which the children regard as naïve and misplaced. Children were quick to challenge assumptions about their innocence, pointing out that they were sexually well informed. Rather than assuming an artificial distinction between the corruption of on-line space and the sanctuary of the home, the children argued that there is nothing which is available on-line which they have not found in pornographic magazines, seen on satellite television or videos at home or heard discussed within their own peer groups.

> Not magazines like *Playboy*, more things like *FHM* and *Loaded*.
> I've never bought *Loaded*.
> Yeah you have.
> Well, Nick, Nick's got em all.
> Nick, Nick buys them all every month, practically subscribes.
> *Loaded*, *Maxim*.
> *Loaded*, *Maxin*.

FHM.

Sky, FHM.

That's not just about making out that's like . . .

Got stories in it.

Yeah.

Position of the week [laughter].

(Highfields boys)

Sexual knowledge and experiences are important currency within teenage culture (Holland *et al.* 1998). Not surprisingly, in all three research areas some boys who were part of particular communities of practice (Wenger 1998) did surf the Web for pornography, representing it as harmless fun. Being interested in computers is usually associated with marginalised forms of masculinity. As Mac an Ghaill (1996) points out, boys who are not interested in traditionally masculine activities such as sport are often labelled as feminine and as socially or sexually undesirable. Yet, by using the Internet to search for stereotypically masculine interests such as pictures of naked women, boys can use ICT to reinforce their understandings of themselves as heterosexual young men.

INTERVIEWER: Have you ever been on pages where you are not meant to be on?

MARTIN: [laughs] Yeah.

INTERVIEWER: What was that, what sort of things?

MARTIN: [laughs]

INTERVIEWER: Don't worry, you're not going to . . .

MARTIN: I've been on porno ones [in very embarrassed laugh].

INTERVIEWER: What at your friends house?

MARTIN: Yeah.

INTERVIEWER: Is it easy to find?

MARTIN: Well it is quite hard cos it keeps saying are you 18 and can you be on this.

INTERVIEWER: And how do you feel when you're on it?

MARTIN: Well it's a laugh really.

Using the technology in this way emerged for these boys as an important way of negotiating their masculinity within the heterosexual economy of peer group social relations (Holloway, Valentine and Bingham in press). In contrast, none of the girls showed interest in seeking out unsuitable materials on-line, generally using ICT in balanced and sophisticated ways, valorising it for schoolwork and for communication.

160

Stranger-dangers

Rare cases of paedophilies using chat rooms to contact children and to set up meetings and even to attempt child abduction have triggered adult fears about children's safety in cyberspace (Lamb 1998; Schwartz 1991). Writing about contemporary stranger-danger moral panics, Oswell argues:

> There has been a shift from a notion of the abuser in the home, which came to prominence in the 1980s, to an older notion of the abuser as stranger, the abuser in the neighbourhood, the abuser in the community. The home and parent are configured as points of safety and security. The Internet is seen as problematic (and dangerous) because it allows the dangerous adult outside to invade the sanctity of that private space. The Internet has now become a focal point for public discussion of child abuse and for shifting the terms of debate away from the home and family as a problematic space.
>
> (Oswell 1998: 143)

Cyberpace is commonly represented as a masculine space which is characterised not only by dangerous individuals but also by abusive and hostile forms of interaction and even rape (Herring 1993). In an article describing her own experiences of being stalked on-line, Pamela Gilbert argues that 'Most of the Net seems perpetually prepared to fight or "flame", at the slightest insinuation. People who would never gratuitously insult others face-to-face are eager to do so in the aggressive environment of the Net' (Gilbert 1996: 126). She goes on to argue that 'The Net is not "just words"' (*ibid.*: 144) but that on-line performances and utterances can have very real material effects for people like herself who have been subject to on-line harassment.

Just as most children are aware of potential stranger-dangers in public space so, too, many of the girls – though few of the boys – recognised the need to be alert to potential dangers in cyberspace. Yet, while acknowledging the possibility that they might encounter what they frequently characterised as 'dirty old men' on-line, the girls also argued that they were not at risk because they were competent and mature enough to take sensible precautions when talking to strangers, to anticipate potential trouble and to avoid putting themselves into dangerous situations.

FRANCESCA: Well it does depend how far you take it doesn't it, cos if you go on and talk to someone and arrange to meet up with them the next day, you know that's taking a risk which you probably shouldn't but. Cos, cos I mean you don't have any personal information on the computer other than say your name, unless like you're talking to someone who's very good at hacking, you know they can't find out where you're from and stuff so in that way I think it's safe, but as I said you get some complete . . .

INTERVIEWER: Weirdo's?

FRANCESCA: Yeah some real, real strange people on there and I mean you know you can avoid talking to them and stuff so it's not too much of a problem, but if you're arranging to meet people and stuff then you really have to know, you have to be sure that they're alright before you do that.

INTERVIEWER: Yeah how do you do that?

FRANCESCA: Um, I don't, I mean there's no foolproof way is there, cos you might be talking to somebody who's a very good liar, but you know the people I chat to, you know I don't know where they are they'd have to be geniuses to make up the kind of lives that they have, you know it's so, it's the kind of details that give it away I think.

As previous work has suggested, children want to be and are very adept at managing their own lives (Alanen 1990). They have a strong sense of invulnerability and often regard their personal safety as their own, not their parents' responsibility (Valentine 1997). Indeed, paradoxically, children often consider their parents to be incompetent both in terms of their understanding of ICT and in terms of their evaluation of the risks it might involve, being too 'panicky' and overprotective.

Time, space and social networks

One of the most popular anxieties peddled about ICT is that these new technologies will have a negative impact on both 'public' and 'private' space, altering children's physical and social relationships to the places in which they play and the homes in which they live (Gumpert and Drucker 1998).

In terms of 'public' space, several commentators have argued that as ICT gives us instant access to non-local people – indeed, sometimes it is faster and easier to communicate on-line than with our physical neighbours – so our social relationships will become liberated from our spatial locality (Meyrowitz 1989). Some writers even go so far as to claim that face-to-face relations will be eroded by on-line simulations, arguing that ICT will displace the need for direct association and human interdependence (Heim 1993). Leary, for example, imagines that in the future:

Face-to-face interactions will be reserved for special, intimate, precious, sacramentalized events. Flesh encounters will be rare and thrilling. In the future each of us will be linked in thrilling cyber-exchanges with

162

many others whom we may never meet in person and who do not speak our phonetic-literal language. Most of our important creations will take place in ScreenLand. Taking off our cyberwear to confront another with naked eyeballs will be a precious personal appearance. And the quality of our 'personal appearances' will be raised to a level of mythic drama.

(Leary 1994: 5, cited in Kitchin 1998: 80)

While Leary's is an apocolyptical vision of the future, other commentators argue that ICT, like television before it (Spiegel 1992), is having a more immediate negative impact on our use of public space, contributing to the contemporary disintegration of the public realm. McCellan (1994: 10), for example, suggests that 'Just as TV produces couch potatoes, so on-line culture creates mouse potatoes, people who hide from real life and spend their whole life goofing off in cyberspace'. Teenage boys, in particular, characterise these representations of techno-addicts who are so absorbed in computers that they become socially isolated loners (Hapnes 1996). It is a representation of childhood that also draws on imaginings of 'the dangerous child' – the computer hacker who can wreak havoc in major computer networks (Elliott 1997). As Oswell explains:

a pathologized image of the solitary working-class boy unsupervised in his television viewing. The image of the boy in a darkened room, with the steely flicker of the television screen on his face is one that is now repeated in popular discourses concerning children's use of computers.

(Oswell 1998: 131–51)

Yet, like adult fears about on-line dangers, children also argued that their anxieties about the effect of on-line activities on their off-line worlds were equally misplaced. Like other technologies, such as the telephone, which was once considered to be an exotic, depersonalising form of contact, but is now regarded as important in sustaining face-to-face relationships and meetings (Fischer 1994), so too ICT generally enhances rather than undermines children's friendships.

New technologies are made sense of by their integration into 'old' ways of life (Morley 1989) so that computer games and surfing the Net have become new interests for children to share with their existing friendships rather than as substitutes for these friends. Notably, the Internet-connected PC plays different roles within different friendship groups and so emerges as a different tool depending on how it is made sense of by different groups of children.

Some children use ICT as an everyday way of maintaining off-line friendships because it is more convenient than meeting face-to-face, especially in Westport, a rural area where the children's homes are very spatially dispersed. Moreover, in contrast to the telephone which generally enables only one person to speak to another, ICT enables several friends to be able to communicate simultaneously. Indeed, using a PC itself is not necessarily a lone activity; rather, some children sit round a screen together to play computer games, to e-mail friends, to participate in chat rooms or to do schoolwork, as Ted describes:

> And, cos before Christmas I think it was, or something, we all, all my friends, well not all of them, about five of us, we, we decided on going to people's houses every week and playing on their computers. And like, some one would bring a play station round and a Nintendo 64, and I'd bring one of my games [computer games] round from upstairs. And cos most of them have got PCs [personal computers]. And then we'd . . . all spend about 3 or 4 hours playing on them.

Even those boys who are branded computer 'geeks' by their peers tend to use PCs in social ways within their own communities of practice to compete at computer games, sharing game cheats and experimenting with the technology in collaboration with other computer enthusiasts (Hapnes 1996). Some mothers whose sons did not fit in at school claimed that computers had saved them from social quarantine by giving them an interest which they could share with other boys who were similarly marginalised by the hegemonic culture predicated on sporting ability.

Those children who frequently use ICT on their own still commonly use their on-line activities as a form of cultural capital which enables them to sustain and develop their existing off-line friendships, as Clive explains below, while other children use ICT to make new on-line friends who share their off-line interests. Such friendships, as in Steve's case below, are often consummated by face-to-face meetings. Rather than turning children into social loners, technology plays an important part in developing, sustaining and maintaining children's off-line and on-line social networks in a diverse range of ways.

INTERVIEWER: Right, so your friends, are those friends that you've made because of the computers and computer games. Or are they friends you had prior . . .

CLIVE: Well one of them, called James, he's he was one I've known, he was in my form all through school. But then, then er a couple of months ago, about the start of this year, both realised we really liked computer games. And that sort of made us more friend like. Or closer friends. And then most of my other friends I just made through computer games. Or computers.

INTERVIEWER: Have you developed any, when you say you chat to people, have you kind of made any friends . . .?

STEVE: Well yeah, I've made quite a few friends and if they all come down like to the park we all, we all have a mess around, we play football and all that.

INTERVIEWER: So how do you, how, like when you're chatting to them on line how do you know they live locally?

STEVE: We don't know, we just like tell them what the address is and they just write back and we ask them if they can send theirs, like Baddow sort of thing and if its says Baddow then we know that they live locally so we can ask them if they can come down one day.

INTERVIEWER: So have people come to visit from other parts?

[edit]

STEVE: Yes around the county.

Like television before it, the PC is accused of displacing the time children used to spend playing outdoors. Yet most of the children interviewed argued that they would rather be outside: hanging on street corners, shopping, at the movies, or playing sport rather than indoors using the computer. Indeed, use of the PC is very seasonal, being largely determined by the weather and number of daylight hours. When it is cold and dark in the winter evenings children prefer to stay indoors and enjoy the freedom which the Internet gives them to communicate with friends and escape other members of the family. When it is warm and light most of the children – with the exception of the most enthusiastic PC users – would prefer to be in public space with their friends. Alistair, and brothers Lionel and Dave explain their balanced approach to the use of ICT:

ALISTAIR: . . . because, like if it's a rainy day I'd like to be indoors playing the computers or whatever. But when it's sunny you want to be out playing sport.

LIONEL: [playing football computer games] . . . it's not as good as actually playing football so we'd think this is a bit boring cos just say they've got one player or something like that and we have to take it in turns and say 20 minutes a go and it'd be better just to play football outside. But depends if it's like a fighting game [on the PC], we're not gonna fight each other or go to the boxing match or something like that.

INTERVIEWER: What about you, Dave?

DAVE: Well I prefer being outside anyway or anything, sort of, you know staying inside so, I don't think it'll really change much except that, I might spend a little longer on the computer than watching TV and anything like that but um, I don't think it'll stop me going outside or anything.

Indeed, ICT often enhances rather than displaces other indoor and outdoor activities. Though 54 per cent of the children we surveyed use a computer at home once a week or more, this does not necessarily mean that their outdoor activities are being undermined. For example, Patrick enjoys making kites and surfs the Net to find out about their history and design, while Carl and his friends enjoying surfing in the sea together just as much as they enjoy surfing the Net together (often to look up sites on surfboards and surfing techniques).

INTERVIEWER: So are the friends, say Leo, friends you use computers with out of school, the same friends you do other things with, like go down the beach and stuff?

CARL: Yeah, it's the same group of friends really. Well I've got a group of friends who are heavily into surfing [in the sea] and really heavily into computers, heavily into like going out and having a laugh down the beach, things.

INTERVIEWER: They're all the same?

CARL: Yeah they're all the same people really.

Other studies (Robinson, Barth and Kohut 1997) suggest that the PC does not necessarily displace other indoor activities, either. People who use new technology are more, not less, likely to use print media and other technologies because books, magazines and the telephone are often used in tandem with a computer, whereas radios, stereo systems or even televisions are commonly used as background accompaniments to working or playing on ICT.

In this way, contrary to the popular moral panics about children and the ICT (outlined in the Introduction), which often imply that children's use of space is dichotomous – they are indoor children or outdoor children, they are on-line children or off-line children – the evidence of our research is that children's use of space, off-line and on-line, outdoors and indoors, is mutually constituted. Children use on-line spaces to find information to help them develop and enjoy their off-line, outdoor hobbies and to make on-line friends who share these off-line, outdoor interests. While playing outdoors, children often talk about and share information they have gathered indoors on-line (for example, tips about how to improve their surfing technique) and develop friends through playing in public space with whom they may also communicate on-line. In this way, technology can shape some children's identities by locating them within particular friendship groups or enabling them to develop particular hobbies. At the same time children can shape the technology by defining and using it in different ways within different peer groups (Valentine and Holloway 1999).

Children's technological competence and domestic relationships

'[F]or the first time in history, children are more comfortable, knowledgeable and literate than their parents about an innovation central to society' (Tapscott 1998: 1–2). They have grown up with computers and have been introduced to them at school, so for them ICT is a normal part of everyday life. Whereas for those adults who have had little or no opportunities to develop technological skills, ICT is regarded as both new and threatening. As a result, some commentators argue that these technologies have the potential to transform traditional social relations and power structures between children and adults (Lumby 1997). Evans and Butkus for example, argue that:

> Cyber-technology has introduced a very different sort of inter-generational wedge; that is, one not constructed by conflicting value systems but by technological competence. Although parents still occupy the role of the initiated with regard to sexuality, if they are uninitiated technologically, then they lose the power base from which to set the markers of progressive socialisation
>
> (Evans and Butkus 1997: 68)

Some children take advantage of this role reversal, teasing their parents about their technological incompetence when they need help. Others are more sympathetic, in some cases providing computer advice for their parents more surrepitiously, and in other cases going so far as to provide structured lessons for their parents to introduce them to new applications (Holloway and Valentine, in press).

INTERVIEWER: How does your Dad feel about having to ask his daughters for help?

RACHEL: Well he doesn't [ask for help], often he has a big panic attack, and he has a big stress and he says 'Urr, can't do it. Stupid computer', and then you say, 'It's only as stupid as you are', and then he says, 'Oh shut up' and things like this and then he doesn't, he won't ask for help straight away like if he can't do it, he'll you know . . .

HELEN: He'll kind of sit there going . . .

RACHEL: . . . he'll progress until he just can't go any further and then he'll say, 'I need some help', and you go, 'Pardon?' and he says, 'Help.' 'What was that?' (laughs) and make him feel really guilty and really bad and he'll just go . . .

HELEN: And then he's there going, 'Help me!'

RACHEL: 'I need some help.' 'Oh help.'

HELEN: And you kinda go, 'Well if you press that', . . . and he's like 'I knew that. I was just testing you'. It's like, 'Yeah, right.'

Although children were often frustrated by their parents' failure to grasp basic skills, and parents were equally frustrated by their children's poor teaching skills and lack of patience, these role reversal relationships are one way that children can renegotiate their relationships with their parents (Holloway and Valentine, in press). Notably, in households where the parents claimed to have open relationships with their children, the children were regarded as socially competent and mature enough to manage their technical competence. In these households, parents did not appear to feel threatened even when their children's technological competence surpassed their own. Rather, by demonstrating maturity through their performance of technological competence, some children were able to enhance their parents' perceptions of their 'adulthood' and negotiate further freedoms. Francesca describes the relationship of trust she has with her parents:

No I think they trust us about that yeah, I mean my Mum sometimes she comes in and she goes, 'What you doing?'. And I go 'I'm just on the Internet', you know 'Are you chatting to people?' I go 'Yeah'. And it's kind of embarrassing cos you feel like she's peering over your shoulder and wondering what you're doing and why are you doing this when you could be out talking to proper people. But they're not, I don't think they're worried, I think we're all, the three of us [herself and two brothers] are kind of too sensible for that.

In contrast, in other homes where the relationships between parents and children were strained or weakened, or where the children had a track record of misbehaving or being untrustworthy (in relation to ICT or other aspects of their lives), parents adopted more hierarchical attitudes towards their children, regarding their technological competence as potentially threatening (Valentine and Holloway, in press). In these households, children's use of ICT appears to be more strictly regulated. Parents use both social and technical means to control their children's technological activities, ranging from direct supervision of their use of ICT to keeping a casual eye on the screen; and from using Net Nannies or passwords to limit children's unsupervised access to ICT, to casually checking the phone bill to monitor use of the Internet. However, parental controls can be quite inefficient, with children finding a range of ways to subvert parental restrictions. Sam describes how his use of the Internet is supervised, while Tim describes his efforts to escape such regulation.

SAM: Mum and Dad usually like supervise me on it to make sure I'm not wasting too much time or anything. I mean I've, I go on a few Web sites of like drum companies and music places, so it's mostly music but I've gone onto find a few, you know, fact things or projects and stuff at school. [edit]

INTERVIEWER: Do you feel kind of constrained about what you can do, I mean would you look at other sites if they [parents] weren't there?

SAM: Well if they weren't there, I mean, because when I go, I used to go on and Dad used to always be over me so that if anything happened or I wanted to do anything he'd do it, so I felt a bit you know constrained as you say, but it's really so that I don't do anything wrong and so I don't waste too much time.

INTERVIEWER: Yeah. Do you ever sneak on it behind his back? [edit]

TIM: Well on like school holidays and me Dad's on afternoons and Mum's working, sometimes I'll sneak on and put a game on. I make sure to cover me tracks.

INTERVIEWER: How do you do that then?

TIM: Well if I've viewed any text things they show up on document on the Start menu bar, so you delete that. And one time it was pretty close. They both went out – a night job, and all of a sudden I heard this car and I went, oh my God! I had this CD on this game and I turned it off straightaway with the CD in and I thought, oh well. And I sat down, (hums), – the TV's not on! Turned the TV on. CD's in! Whoosh! [Sound of taking CD out] And they just came in and I went, "Hi!"

INTERVIEWER: I'm watching this really interesting documentary.

TIM: I think me Mum forgot something when she came in. I went, bye. I thought now what have I done to this. It were all right actually. Cos sometimes if you switch it off with CD in it damages it, but I managed to put it on and delete it back off. It were pretty close.

In this way, ICT is a tool some parents use to exercise care and control over their children while simultaneously also being a tool some children use in negotiating their autonomy and independence from their parents. In other words, ICT can play a range of roles within different households and so emerges as a different tool in different families depending on the different ways it is made sense of by them. This is a reciprocal process of change; while parents' perceptions of children's social competence shapes the way their technological competence is allowed to develop, the children's technological competence can also shape relations in the family. In other words, social and technological competencies co-develop (Valentine and Holloway, in press).

Conclusion

Adult moral panics about children's use of computers assume that technology will impact on children's lives in negative ways: corrupting their innocence, potentially exposing them to dangerous strangers on-line, undermining their friendships and use of space, and eroding adults' authority over them. These moral panics are not only technologically determinist, in that they assume that technology will have particular outcomes regardless of the complexity of children's relationships with media or the contextual relationships within which these are situated, but they are also adultist in that they make assumptions about children's practices. Children are presupposed to be less able than adults to distinguish between suitable and unsuitable sites; less able than adults to handle potential on-line dangers; and less capable than adults to use technology sensibly without becoming techno-addicts and social loners.

Yet the evidence of this research is that, contrary to these moral panics, computers cannot be viewed as invariant objects, nor as impacting on children's lives in fixed ways producing a predictable set of ill-effects as a result of some internal technological logic. Rather, the properties of ICT emerge in practice as they are domesticated by individuals and households in what is a reciprocal process of change (Bingham, Holloway and Valentine in press; Law 1994).

Although technology can shape some children's identities (for example by binding them together in particular friendship groups or by enabling them to perform their competence in such a way that it alters their parents' understandings of their maturity), the technology itself can also be transformed according to the interests and meanings of individuals and groups of users, such that it emerges as a different tool for different groups of children (for example some valorise ICT as a tool on which to play computer games with off-line friends, while others valorise it as a tool of communication, using it to establish global on-line friendships).

As a result, this chapter exposes the inaccuracy of adultist assumptions about children's use of technology. Though adult moral panics appear to assume a binary distinction between on-line and off-line worlds, children implicitly recognise that these spaces are not separate but are mutually constituted. Specifically, they point out that pornography and other unsuitable materials are not unique to cyberspace, but are readily available in off-line enviroments. Moreover, the everyday social practices of children described in this chapter show that children who use Internet-connected PCs are not socially isolated indoor children, but rather that they use on-line spaces to develop their off-line hobbies, use of space and friendships. Thus, contrary to popular fears, children tend to use ICT in balanced and sophisticated ways.

Acknowledgements

Gill Valentine and Sarah Holloway wish to acknowledge the support of the Economic and Social Research Council for funding the research on which this paper is based (award no. L129 25 1055).

References

Alanen, L. (1990) 'Rethinking socialisation, the family and childhood', *Sociological Studies of Child Development* 3: 13–28.

Aune, M. (1996) 'The computer in everyday life: patterns of domestication of a new technology', in M. Lie and K.H. Sorensen (eds) *Making Technology Our Own?*, Oslo: Scandinavian University Press.

Bingham, N., Holloway, S.L. and Valentine, G. (in press) 'Bodies in the midst of things: re-locating children's use of the Internet', in N. Watson (ed.) *Reformulating Bodies*, Basingstoke: MacMillan.

Bromley, H. (1997) 'The social chicken and the technological egg: education, computing and the technology/society divide', *Educational Theory* 47, 1: 51–65.

Bryson, M. and de Castells, S. (1994) 'Telling tales out of school: modernist, critical and post-modern "true stories" about educational computing', *Journal of Educational Computing Research* 10: 199–221.

Cate, F.H. (1996) 'Cybersex: regulating sexually explicit expression on the Internet', *Behavioral Sciences and the Law* 14: 145–66.

Elliott, C. (1997) 'Found: spy who hacked into Pentagon during A levels', *Guardian*, 22 March, p. 1.

Evans, M. and Butkus, C. (1997) 'Regulating the emergent: cyberporn and the traditional media', *Media International Australia* 85: 62–9.

Fischer, C. (1994) *America Calling: A Social History of the Telephone to 1940*, Berkeley, CA: University of California Press.

Gilbert, M. (1996) 'On space, sex and being stalked', *Women and Performance: A Journal of Feminist Theory* 9, 17: 125–50.

Gumpert, G. and Drucker, S. (1998) 'The mediated home in the global village', *Communications Research* 25, 4: 422–38.

Haddon, L. (1992) 'Explaining ICT consumption: the case of the home computer', in R. Silverstone and E. Hirsch (eds) *Consuming Technologies*, London: Routledge.

Hapnes, T. (1996) 'Not in their machines: how hackers transform computers into subcultural artefacts', in M. Lie and K.H. Sorensen (eds) *Making Technology Our Own?*, Oslo: Scandinavian University Press.

Heim, M. (1993) *The Metaphysics of Virtual Reality*, New York: Oxford University Press.

Herring, S. (1993) 'Gender and democracy in computer mediated communication', *Electronic Journal of Communication* 3, 2: 221–36.

Holland, J., Ramazanoglu, C., Sharpe, S. and Thomson, R. (1998) *The Male in the Head: Young People, Heterosexuality and Power*, London: Tufnell Press.

Holloway, S.L. and Valentine, G. (in press) '"It's only as stupid as you are": children's negotiations of technological competence at home and school', *Social and Cultural Geography*.

Holloway, S.L., Valentine, G. and Bingham, N. (in press) 'Institutionalising technologies: masculinities, femininities and the heterosexual economy of the IT classroom', *Environment and Planning A*.

Kitchin, R. (1998) *Cyberspace*, Chichester: John Wiley & Son.

Lamb, M. (1998) 'Cybersex: research notes on the characteristics of the visitors to on-line chat rooms', *Deviant Behaviour: An Interdisciplinary Journal* 19: 121–35.

Law, J. (1994) *Organising Modernity*, Oxford: Blackwell.

Leary, T. (1994) 'How I became an amphibian', in *Chaos and Cyberculture*, Berkeley, CA: Ronin.

Lumby, C. (1997) 'Panic attacks: old fears in a new media era', *Media International Australia* 85: 40–6.

Mac an Ghaill, M. (1996) 'Deconstructing heterosexualities within school arenas', *Curriculum Studies* 4: 191–209.

Markoff, J. (1995) 'On-line service blocks access to topics called pornographic', *New York Times*, 29 December, p. A1.

McCellan , J. (1994) 'Netsurfers', *Observer*, 13 February.

McRobbie, A. and Thornton, S. (1995) 'Rethinking "moral panic" for multi-mediated social worlds', *British Journal of Sociology* 46, 4: 559–74.

Meyrowitz, M. (1989) 'The generalized elsewhere', *Critical Studies in Mass Communications* 6, 3: 330.

Mitchell, D. (1995) *City of Bits*, Cambridge, MA: MIT Press.

Morley, D. (1989) 'Where the global meets the local: notes from the sitting room', *Screen* 32, 1: 3–15.

Murdock, G., Hartman, P. and Gray, P. (1992) 'Contextualising home computing: resources and practices', in R. Silverstone and E. Hirsch (eds) *Consuming Technologies: media and information in domestic space*, London: Routledge.

Nixon, H. (1998) 'Fun and games are serious business', in J. Sefton-Green (ed.) *Digital Diversions: youth culture in the age of the multi-media*, London: UCL Press.

Oswell, D. (1998) 'The place of "childhood" in Internet content regulation: a case study of policy in the UK', *International Journal of Cultural Studies* 1, 1: 131–51.

Prout, A. and James, A. (1990) 'A new paradigm for the sociology of childhood? Provenance, promise and problems', in A. James and A. Prout (eds) *Constructing and Reconstructing Childhood: contemporary issues in the sociological study of childhood*, Basingstoke: Falmer Press.

Robinson, J.P., Barth, K. and Kohut, A. (1997) 'Personal computers, mass media and use of time', *Social Science Computer Review* 15, 1: 65–82.

Schwartz, J. (1991) 'Sex crimes on your screen?', *Newsweek*, 22 December, p. 66.

Silverstone, R., Hirsch, E. and Morley, D. (1992) 'Information and communication technologies and the moral economy of the household', in R. Silverstone and E. Hirsch (eds) *Consuming Technologies: media and information in domestic space*, London: Routledge.

Spiegel, L. (1992) *Make Room for TV*, Chicago: University of Chicago Press.

Squire, S. (1996) 'Re-territorialising knowledge(s): electronic spaces and virtual geographies', *Area* 28, 101–3.

Tapscott, D. (1998) *Growing Up Digital: The Rise of the Net Generation*, New York: McGraw Hill.

UK Government, 'IT for All' survey, <http://www.itforall.gov.uk/it/survey/3.html>.

Valentine, G. (1996) 'Angels and devils: moral landscapes of childhood', *Environment and Planning: Society and Space* 14: 581–99.

—— (1997) '"Oh yes I can." "Oh no you can't." Children and parents' understandings of kids' competence to negotiate public space safely', *Antipode* 29, 1: 65–89.

Valentine, G. and Holloway, S.L. (in press) 'Virtual dangers? Geographies of parents' fears for children's safety in cyberspace', *Professional Geographer*.

—— (1999) 'Cyberkids?: exploring children's identities and social networks in on-line and off-line worlds', paper available from the author, Dept. of Geography, University of Sheffield, Sheffield, UK.

Vanderkay, J. and Blumenthal, S. 'CI find that more than half of households with children have PCs', Rogers Communications, <http://www.ci.zd.com/news/ctic97.html>.

Vestby, G.M. (1996) 'Technologies of autonomy? Parenthood in contemporary "modern times"', in M. Lie and K.H. Sorensen (eds) *Making Technology Our Own?*, Oslo: Scandinavian University Press.

Waksler, F. (1986) 'Studying children: phenomenological insights', *Human Studies* 8: 171–82.

Wenger, E. (1998) *Communities of Practice*, Cambridge, Cambridge University Press.

Wilkinson, H. (1995) 'Take care in cyberspace, children', *Independent*, 1 December, p. 21.

10

YOUNG CARERS IN
SOUTHERN AFRICA

Exploring stories from Zimbabwean
secondary school students

Elsbeth Robson and Nicola Ansell

Introduction

Children and young people throughout the world have always been involved with caring for others, whether elderly, ill or disabled parents, siblings or other family, household or community members. In the developed world the hidden minority of children engaged intensively in such activity have recently been 'discovered' and labelled 'young carers'. Since the mid-1980s, by identifying and defining young carers, a set of advocacy institutions and a pool of research literature have evolved, mostly concerned with the UK (Aldridge and Becker 1993; Aldridge and Becker 1995; Becker, Aldridge and Dearden 1998; Segal and Simkins 1993). By highlighting the neglected situation of young carers (who until recently very often fell through the welfare net of social services), researchers and social work practitioners generated a great deal of media and policy interest. So far the focus geographically has been on young carers in the North[1] (Becker *et al.* 1998). This exclusivity led us to ask, 'What of young carers in the South?'

Young people as carers have not been a research focus in the South because, in many societies of the developing world, caring by children is constructed by anthropologists, social scientists and others, as part of socialisation, something expected of children, a normal part of growing up. This may reflect common practice, but not all such 'cultural' practices should be accepted uncritically. It is necessary to explore the impacts caring has on the lives of young people within these changing societies.

Equally, it would be inappropriate to see all involvement of children in 'work' as exploitative, as is sometimes implied by international development bodies seeking to impose normative notions of childhood, developed in the North, on very different social and cultural contexts in the South. Caring should neither be simply romanticised as part of unchanging traditional patterns of behaviour, nor condemned at the other extreme as exploitation. Rather, it needs to be acknowledged that caring for a disabled, elderly or sick household member may impose particular and at times challenging demands on young people and, as such, these children may be considered 'children in especially difficult circumstances' (UNICEF's term; see Gwaunza *et al.* 1994) who deserve greater attention and possibly supportive action.

Zimbabwe is the regional focus of the empirical research recounted in this chapter. It is to be expected that significant numbers of young people in the country find themselves in caring situations because of socio-cultural expectations, poor or non-existent social services, poverty and AIDS. Despite their junior household status, children in Zimbabwe do much unpaid work (Reynolds 1991). Their roles as unpaid workers looking after siblings, tending livestock, working alongside adults in domestic reproduction and farm production reflect prevailing socio-cultural constructions of childhood here (Gelfand 1979). That Zimbabwe's young people work without pay also reflects the level of poverty in the country, where 68 per cent of the population in 1990–1 were living on under US $2 a day and 26 per cent were below the national poverty line (World Bank 1999: 197). Zimbabwe is currently in a period of economic decline and since 1990 has been undergoing the rigours of economic structural adjustment policies. These mean that health and other welfare services are being squeezed, with increasing burdens on the household (Renfrew 1996). AIDS in Zimbabwe is having serious and growing socio-economic impacts. Current estimates suggest that 20 to 26 per cent of people aged 15–49 live with HIV or AIDS (UNAIDS 1998). By 1996, 8 per cent of Zimbabwean children under 15 were maternally orphaned by AIDS, a figure set to rise to 16–22 per cent by 2001 (Foster *et al.* 1997: 2). Thus, in Zimbabwe, young people increasingly find themselves as carers because children are expected to help at home, AIDS deaths among adults and numbers of sick adults with HIV–AIDS are growing, state healthcare provision is being eroded and increasingly impoverished households can less afford to purchase healthcare.

In this chapter we analyse some accounts of caring by rural young Zimbabweans, asking 'What do these stories tell us about young people's experiences as carers in Zimbabwe?'

Context

The school

The stories analysed in this chapter were written by students at Ruchera[2] Secondary School, Manicaland. The school is situated in a communal farming area of eastern Zimbabwe about 200 km from the capital Harare and 33 km from the main Harare–Mutare road (Map 10.1). It is a fairly 'typical' rural school in most respects, although the surrounding community is slightly more affluent, enjoying better amenities than many of Zimbabwe's rural areas. The school is located 2 km from the rural business centre of Chewere, from which there are frequent buses to Harare, a journey of three or four hours. The school is a District Council day school constructed by the local community and has approximately 500 students and 22 teachers. Like the surrounding community, the school has no electricity or telephone (unlike the business centre) and water is drawn from a borehole 200 metres from the campus. The teachers consider the school to have better facilities than some rural schools, but far worse than mission-run boarding schools, or most government schools. Although it is not

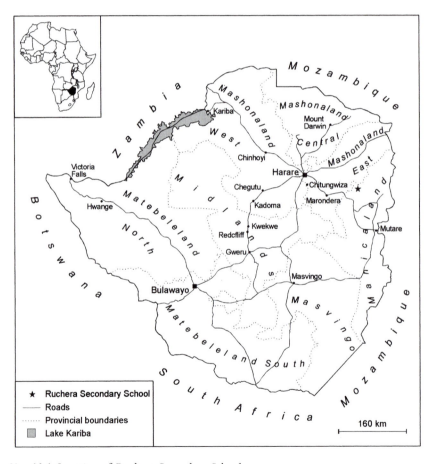

Map 10.1 Location of Ruchera Secondary School

a mission-run school, like most schools in Zimbabwe, Ruchera Secondary School has a Christian ethos expressed in the religious content of assemblies, religious education lessons and the activities of the scripture union.

The young people

Ruchera's students are drawn from the local catchment, walking, cycling or bussing daily up to 12 km each way from their homes. Some have parents in urban areas, but reside with grandparents near Ruchera in order to take advantage of the lower rural school fees and to have a rural upbringing, as many parents appreciate the 'traditional' values thereby instilled in young people. Historically among the Shona, children were sent to stay with their grandparents to learn how to behave properly and to provide help (Gelfand 1979). Many students belong to households headed by men resident elsewhere. A few students live in households with no resident adult, as their parents are employed elsewhere, or have died (usually of AIDS).

Settlement in the school's catchment is dispersed, with no clear boundaries between homesteads of neighbouring villages. Students generally belong to farming households whose livelihoods depend on male migrant labour as the most important source of income, followed by farming. Maize is cultivated (both for subsistence and as a cash crop), as well as beans, groundnuts and a variety of vegetables for both household consumption and small-scale local trading. Livestock-keeping by about a quarter of households is a further, but relatively minor, livelihood source. The majority of students live in homesteads constructed of locally baked bricks, most with latrines, and about half have their own household borehole. Although there are severe inequalities between households, many possess capital goods such as wheelbarrows, ox carts, and sewing machines. A few households have more expensive items such as bicycles and televisions (powered by solar panels).

The students at Ruchera Secondary School belong to a fairly narrow socioeconomic group. They do not come from the poorest households in the vicinity – in this particular rural area about half the young people of their age do not make it into secondary school at all, so the students tend to belong to the slightly better off, but not the wealthiest, rural households. Within those households only some young people will pursue secondary education according to their ability, parental choice and, to some extent, their own inclinations. These factors partially explain the skewed gender balance of the school, whose enrolment comprises 43 per cent girls, 57 per cent boys.

The students share a common ethnic background, all being Shona-speaking (Zimbabwe's biggest language group). The students in forms 2A and 2B who wrote the stories analysed here range in age from 14 to 17 years and have nine years of schooling (seven years primary school, two years secondary school), of which all but the first three years of primary school are largely English medium. The students share a common religious background as, like most Zimbabweans, they would describe themselves as Christians, and most attend church on Sundays.

The research

The students' stories of experiences as carers were collected in September 1997 by Nicola Ansell, who had been present in the school as an outside researcher for two months and was familiar to the students. The research took place during a lesson period when the usual teacher was absent, and was organised in a manner which conformed to students' expectations of a lesson format. The session began with an open discussion of caring: students were asked to describe occasions on which they had helped elderly, sick or disabled people and to tell the class who they had helped, what help they had given, and how they felt about it. They were then asked to write a composition, in English, with the title 'Caring for other people'. The composition was written by all students present that day (44 in form 2A and 37 in form 2B).[3]

Using stories in this way is effective for gathering information in students' own words about practices of caring, as well as students' attitudes to their roles. Students are able to recount their experiences in their own terms, conveying as much, or as little, of their feelings and thoughts as they wish, in circumstances less intimidating than face-to-face interviews.

However, the essay format does not allow for total freedom of expression. The first constraint which needs to be noted is the fact that the compositions were written in English, these young people's second language. Though the students are relatively proficient in English, it is a language in which they are used to expressing only certain types of thought. Furthermore, their English vocabulary is limited, restricting their range of expression. In writing, subtle nuances are seldom available to them, or may be misinterpreted by the reader. At the same time, it should be recognised that the students are more familiar with writing in English than in Shona, and it is conceivably more empowering to allow young people to express their thoughts in English, rather than subjecting them to another layer of interpretation from Shona.

Language, however, is by no means the only constraint the research method imposed on the young people. The set-up of the exercise, the provision of title and guidelines (instructions to write about experiences of caring for the ill, very old or disabled; members of the family, relatives and neighbours) imposed expectations. Suggestions made by students orally at the beginning of the lesson are also likely to have influenced the others in their efforts to provide what was expected – the usual purpose of students in writing classroom compositions.

Even the form of the composition is a potential constraint – expectations concerning length, the degree of self-revelation, the structure of the narrative. All these factors demand that the compositions be read, not at face value, but as texts and as forms of discourse.[4]

It should also be recognised that not all students have experience of caring for others in a practical way which they can easily write about. One feature of the use of story-writing as a method, is that students are able to elaborate on their experiences, or produce entirely fictitious accounts, drawing on their imaginations, as well as on wider discourses of caring, health and old-age prevalent in the society in which they live. Thus, not only do the compositions comprise a discursive form in themselves, but they need to be read in relation to, and help us to understand, the discourses which inform young people's actions in relation to caring, and which they mobilise in making sense of their own experiences.

It became apparent, in reading the compositions, that they generally could be classed into one of two categories. The majority were clearly either fictional narratives describing helping people in only very general terms, or related incidents such as helping an old person to carry their shopping. These essays are analysed primarily for their potential to illuminate general discourses of caring, and their contribution to interpreting the stories told by students with more 'real' caring experiences. However, eighteen essays appeared (not least by containing strong and specific details) to recount actual caring experiences. Of this subset, only nine cases are analysed in depth here. These were selected as the stories relating the most real and detailed experiences of caring. This is not to suggest that they are unmediated accounts of the experience of caring, from which 'reality' can unproblematically be reconstructed. All of the accounts undoubtedly contain elements of fantasy and exaggeration, as well as the distortions of memory and selective reporting.

However, the 'real' narratives of caring stand out as including very specific or personal details, which convey the powerful reality of caring experiences for the young people themselves. For example, Charles (16-year-old boy) relates the precise time period and dates of caring: 'It was already last year in September I helped my grandfather who was very ill . . . until his death on 14th December 96' (Box 1). Similarly, some of the young people convey vivid details of the caring experience, for example Nancy (15-year-old girl) writes of washing her elderly sick grandmother's clothes and notes that 'some of the clothes were mercy (mucky) with her vomitters (vomit)'.[5] The stories are analysed to elucidate the characteristics of the youngsters and care recipients, and the care work performed and young people's responses to their caring experiences, as well as the ways in which more general discourses of caring are mobilised by students in making sense of their experiences.

Box 1 Charles's Caring Story

The time I helped a person who was ill

It was already last year in September I helped my grandfather who was very ill. The first days I thought that if I help him I will be affected by such a disease. At that time he was about fifty year old, but he looks like a young.

As I saw that disease does not want to finished I took him and the went to the hospital to be vaccinated. The doctors gave him some vaccines, but there is no change. I helped him in many ways.

I went to his home and I cooked delicious meal every day. His clothes I took and washed them. Sometimes I gave him the vaccines that he had given to the hospital. After each supper I gave him some vaccines and I go to prepare him to slept.

The first day I thought that I will affected by such a disease, but I secrified to help him in many ways. I help him until his death on 14th December 96.

Charles Marume, 16-year-old boy, 18 September 1997

Children's 'real' and 'imagined' caring experiences

Who cares?

Most studies conducted among carers in the North have found the majority to be female (Becker *et al.* 1998: 34–5; Brown and Smith 1993; Glazer 1990; Gregor 1997), a finding replicated in research among young carers (Becker *et al.* 1998: 23). It is not, then, surprising that girls are over-represented in terms of the reporting of 'real' caring experiences by young Zimbabweans. Though the classes were dominated by boys (50 of the 81 students), the majority of those reporting actual experiences of caring were girls (10, compared with 8 boys). This mirrors findings elsewhere in Zimbabwe that women and girls dominate caring reflects the dominant social construction of caring as female work (Robson 1998). What is more unusual is that the proportion of boys telling stories of periods of intense caring (see, for example, Box 2) is relatively high. The large proportion of boys in the sample makes this an unusual, and especially valuable, set of cases of caring by young people, providing particularly useful insights into boys' caring work, which is generally more invisible even than girls' caring work.

Box 2 Loveless's caring story

In 1994 when I was 14 year old. My grandfather be come ill and he was not able to stand; eat and bath. He was 70 year old in that year.

He was having a problem in his lags.

When he was ill a chose to sleep with him and bath him and all – so find a car to carry him to the hospital. When he was there I chose to stay their antill he was well. But it took about three coter's of a year. I was also chose to clean his waste product.

When I was helped him I was feel to find some doctor and nases to help him and I was feel angry because I was one who was help him. I feel to find some madisan to care of a disease like BP. Becoase he was suffering with that disease.

Loveless Chigondo, 17-year-old boy, 18 September 1997

Interestingly, it appears that not only do boys find themselves in situations in which they are providing care for other people, but they have little difficulty imagining themselves in such situations. The students' 'fictional' responses differed little between girls and boys, and there was no evident reluctance of boys to situate themselves in caring roles.

Who is cared for?

We can begin to understand why boys are so involved in caring by reference to the gender of the care recipients. In nearly all cases of 'actual' care provision there was gender matching of the carer and care recipient. Socially and culturally, such gender matching is important. Research in the North indicates that, particularly where intimate care (e.g. bathing and toileting) is involved, a girl caring for a man, or a boy caring for a woman, is considered unacceptable (Becker *et al*. 1998: 20), but may be unavoidable in some circumstances. Historically, intimacy between members of the opposite sex outside marriage beyond puberty is strictly prohibited in Shona society, there being an implicit belief that this is likely to lead to sexual intimacy. Strict taboos against incest mean that contact within the family is particularly avoided (Gelfand 1979), making caring across genders unacceptable. In the only case of unmatched gender between carer and care recipient, intimate care is not reported (Box 3).

Among the 'fictional' accounts, a slightly different picture emerges. A number of the girls (although a minority) imagined themselves caring for elderly or disabled men (although this did not extend to descriptions of providing intimate care). Among the boys, however, no more than two or three responses suggested that they might provide care of any form to a woman. It was almost invariably elderly, sick or disabled men who they reported finding hauling heavy bags, or lying at the roadside in need of help. It is clear that boys do not expect to find themselves in a caring role *vis-à-vis* women, which may again reflect sexual taboos, or may simply be considered an affront to male dignity.

The type of person children portrayed themselves helping in their fictional accounts, is a stranger, or someone known only distantly, maybe a neighbour, but almost never a family member (and never a close relative). This perhaps reflects a refusal to contemplate illness, or disability, in the young person's immediate family, but might also be reflective of the biblical narratives young people appear to draw on in relating stories of caring. Many of the stories bear a very close resemblance to the parable of the Good Samaritan, the young person happening upon a person in need of help while out shopping or walking, providing immediate help, and perhaps visiting the person after a few days. These stories differ markedly from the experiences recounted with a more convincing ring.

Among the 'true' stories, grandparents figure highly (five of the nine cases) as care recipients. Taurai, for example, cares for both his grandparents whom he lives with (Box 3). In some cases young people's care for the frail elderly extends beyond the immediate household. Thus, Norman (15-year-old boy) cares for an elderly neighbour whom he refers to in his composition variously as 'a old grandfather', 'our neighbour' and 'our grandfather'. Unlike in the fictional accounts, however, this elderly man was clearly well known to the family, and care was ongoing.

Not all the grandparents appearing in the stories needed care because of the frailty and impairment of old age. Charles's grandfather 'was about fifty year old, but he looks like a young' when Charles helped care for him until he died (Box 1). The young people also recount caring for other family members including a mother (Box 5), an uncle and an aunt. Once again, these are close relatives who do not figure in young people's more general discourse of caring. Circumstances in practice do not coincide with imagined situations.

According to the young people's stories, the care recipients needed care for various reasons. In three cases they were elderly and frail, as Norman describes: 'Last year on November I helped a old grandfather. At this time he is 95 years old. He can not walk properly he is using a stick . . . When I cooked he super he need soft meat because he don't have teeth.' 'Heart cancer', 'ulcers' and 'blood pressure' were each named as (at least part of) a care recipient's terminal illness. These reasons for care differ markedly from those of the students' imagined fictional accounts, which tended to relate to mobility difficulties, blindness

Box 3 Taurai's caring story

. . . I want to tell how to care people are very old. I live with my grand-parents because my parents live in Harare where the works my father is an engineer whilst my mother is a teacher. I help to care my grandparents by cooking and preparing food to eat, to the borehole and fetch water to sweep the yard and house, to herd cattle, sheep, goat and pig . . .

Taurai Tapfuma, 15-year-old boy, 16 September 1997

and mental illness. This raises questions concerning the extent to which young people are able to mobilise the more general discourses of caring in making sense of their 'real' experiences.

However, a common thread running through the actual and imagined caring stories is that all of the students reproduced a discourse which constructs care recipients as victims, incapable of caring for themselves, and deserving of pity. For example, Sungai (16-year-old boy), writing in very general terms, says of disabled people, 'To thise people, I feel compassion to them. I see them as sheep without a sherperd', again employing a biblical analogy. Similarly Netsai (14-year-old girl), who had helped her mother recovering from an accident, explains 'I feel pity of disabled people because there are not the ones who choosed to be a criminal.' Choice of the word 'criminal' perhaps reflects Netsai's limited English vocabulary, but is nonetheless indicative of the lack of respect felt towards people with disabilities, and the attitude that they are a burden on society.

How do young people care?

In their stories the young people reported practical caring in a variety of ways. They undertook various domestic and farm work tasks, either directly generated by the care recipient's illness (e.g. washing soiled clothes and linen), or because the illness left them unable to work. As Patikayi (14-year-old boy) explains, 'I myself helped my uncle too by harvesting his crops, washing his clothes, feeding his cattle and all the thing that deams necessary for him'. Much of the work reported is the same as work that young people normally undertake in rural Zimbabwean homes, whether living with their parents, grandparents, or alone. Daily diaries from other students in the school report doing similar tasks as students with caring stories. These tasks fall into the categories of domestic chores (cooking, washing clothes, fetching water, heating water on the fire, sweeping, shopping, home gardening), which are carried out more by girls than by boys, and farm work, done almost exclusively by boys (including harvesting crops, feeding and herding cattle, sheep, goat and pigs). The gender divisions largely reflect wider gendered constructions of domestic labour in what can be loosely termed 'Shona culture', in which caring for people and gardens is largely female-specific, but caring for animals is largely male.

However, young people's normal daily chores do not include giving adults direct personal care, helping with daily living and providing hospital-related care. Undertaking these tasks may be considered to mark out particular 'caring' experiences, as distinct from the routine domestic work of young people.[6] Help with daily living reported includes helping the care recipient to walk and to make their bed. When a household member is hospitalised, or attends a medical centre for treatment, an additional burden of care falls on the household. So young people are frequently required to accompany the care recipient for treatment, take food to them daily (in African hospitals, patients' families generally have responsibility for feeding) and even stay with them in hospital. Loveless provided hospital-related care when his grandfather was ill, including seeking transport to get to the hospital and staying there with him (Box 2). These are the types of tasks that most students imagine performing for those who are sick or disabled.

For many Northern carers and their families, provision of personal care is the most distressing and unacceptable face of young caring (Becker et al. 1998: 20). The Zimbabwean students report intimate caring in their 'true' stories, including sleeping in with the care recipient, feeding, bathing and toileting (see Box 4). They also provide more general personal care, such as giving medication, comforting and encouraging. For example, Netsai talks about the importance of providing reassurance while looking after her mother: 'while she is on bed, giving her strength words so that she will not think that she about to die'.

Box 4 Mugowe's Caring Story

It is a good thing to help people who are ill disabled, old, family, friends and neighbours if they are in troubles.

I help my grandfather last year who was ill he took about 6 months without doing anything to help himself but I help him untill his death.

I help him to go to the toilet because he was not to walk and I help him to make his bed. I gave him food.

I feel sorry about it because it is not a good thing to see other in occasions like that.

Mugowe Marume, 17-year-boy, 16 September 1997

In almost all cases the young people were carrying out caring in their own home, which for Taurai is his grandparents' household where he lives. In only two cases did students go to care for an individual in another homestead: Charles went to care for his grandfather in his grandfather's own home (Box 1), while Norman cared for an elderly neighbour in a homestead nearby.

Almost all the care experiences young people reported were from the past and, where specified, were mostly from the recent past, that is the past one to three years. This could indicate children as young as 12 taking on caring responsibilities. Only Taurai, caring for his grandparents with whom he lives, was explicitly involved in ongoing caring. The periods of caring reported extended between two weeks and nine months. In about a third of the cases caring terminated with the care recipient's death (not an outcome anticipated in the fictional accounts). In only two cases was the care recipient reported to recover – one from illness and another from an accident. Either way, in most cases caring was temporary, which suggests it may form an interval (if not so short as those in students' 'fictional' accounts), rather than a permanent feature, in young people's lives.[7]

Why do young people care?

Having examined in these stories the who, what, where and when of young people's caring, we now try to answer why? Why do these young people, like other young people in Zimbabwe and elsewhere, find themselves caring, or choose to be carers? In about half the 'true' stories young people do not give specific reasons why they care, but appear to care willingly, from choice or sense of duty. There are strong cultural expectations that young people undertake tasks within the home which extend to caring for others.

Though most of the young people, in their accounts, present caring as an option, in both 'fictional' and 'actual' stories, it is also presented as a duty which a good Christian should perform whenever required. Students mobilise discourses which promise rewards for caring, although it is made clear that this should not be the primary motivation. In particular, several of the young people in their fictional accounts tell of being rewarded with cash, which most of them reject: 'I saw the man find money and I told the man that I was helping only don't give me money although I know that money is the root of all evils' (Garfield, boy, unspecified age). More prominent were the assertions that caring is something which brings blessings from God, and admiration from people. 'When I help some other people I feel well becouse if you help anothe person God will give you talents and some ather people will love you' (Munodeyi, 16-year-old boy). Thus, students borrow symbols and phrases from the Bible in their narratives.

Though these are the broader explanations used by young people to explain their involvement in caring activities, it is also appropriate to look at the more concrete circumstances in which young people find themselves, and the ways in which they understand their caring roles to have come about. Clearly, Taurai (Box 3) is under the authority of his parents, who dictate that he live with his rural grandparents to attend school – a situation that involves also caring for his grandparents. However, there is also evidence for some degree of autonomous decision-making by young people themselves about whether, or not, to care. For example, at times Loveless says he chose to care for his grandfather and at others that he was chosen to care (Box 2).[8]

Box 5 Stella's caring story

Some people take it as something very disgusting like clearing someone vomiting, spitting and other things. To me I think it is something that is very simple, you don't know what is coming to you tommorow maybe it might be you who will be sick. Its just the same as do unto others as you would have them do unto you.

I have my mothers sister who was suffering from alsers. But she died last year. Everytime there was nobody around, I would cook for her, go to the shops, heat her water inorder for her to take a bath. Even though she used to vomit, spit, I didn't mind I would just wash the clothes like there was nothing and I would take it as my own thing.

Well by helping her and she would thank me my tears would drop and I felt there was no need for her to say so. I would feel really happy for what I did and you know its something you should be proud of to see someone you helped well again. That's what I also wanted, to see her walk again but unfortunately she died anyway its part and parcel of life.

She didn't used to eat food with curry so I would try sometimes buy her bananas, oranges apples, pine apples and it would fill my heart with joy to see her eating and smilling. again. I really felt good bout it but some people laugh at other people who are sick or injured but to me its natural I really feel pity like its my own brother, sister or parent. Its something thats just natural to me.

In the streets sometimes I pass by you see people begging for money. Their clothes dirty, torn and I really feel bad about it. If possible I give them money. It sometimes make me wish if I had a house and I would collect them and stay with them all. I really feel like they are part of my life. But when I tell other people they don't understand me they say I'm strange but even though it doesn't change my heart or turn it around its just always the same.

Stella Chawatama, 14-year-old girl, 16 September 1997

Some students made comments that give greater insight into why young people in particular situations care. Stella explains that she took care of her aunt 'Everytime there was nobody around' (Box 5). She is clearly suggesting that if 'somebody' is around, presumably an adult, the task of caring for her aunt would probably fall to them. She regards her own role as one of substituting for an adult carer. It has been shown that, in a myriad of different circumstances, young people care because there is no available adult for the task. Potential adult carers may not be present in the household – they may have died, be absent migrant workers, mentally unstable or even unwilling to do the work. Caring by a young person would appear to be seen as second best. This must have implications for how young people see themselves.

Young carers may also understand their situation in relation to their position within the family. Where there are siblings, the burden of care may fall on one child rather than others. Netsai told us that when her mother needed care, 'I her elder child was the one who was responsible for her'. Which child, of siblings, becomes the carer is usually an outcome of various factors including their relative ages, gender and aptitudes, as well as gender of the care recipient and the household livelihood situation. This reflects and helps constitute discourses concerning age and gender. What can be gleaned from these accounts about why children become carers confirms the factors identified by Becker et al. (1998: 22), which include nature of the care recipient's illness or disability, household structure, gender, co-residence, status, power and degree of external support. However, it does not necessarily follow that household decisions about caring are made on the same bases in all families, let alone in the different settings of Zimbabwe and the UK.

Responses to caring

It has already been suggested that young people make sense of their involvement in care-giving in relation to broadly Christian discourses of morality that construct caring as a duty, and something to enter into willingly and without complaint. It should be acknowledged here that a characteristic of the use of storytelling as a methodology is the silence on the subject of negative feelings and resistance. Given the normative notions of caring imbuing the discourse, it is hardly surprising that no stories reveal strongly negative feelings directed either towards caring tasks, or towards care recipients. Young people resisting caring may resort to passive non-cooperation, or in extreme situations leave the household to live with other kin, or join the growing number of urban street children; such strategies are unlikely to appear in their stories.

The narrative structure employed by most students emphasises the duty to care, and benefits accruing from performing such duties. Many of the stories, both 'factual' and 'fictitious', open with a short section along the lines of 'It is good to care for the sick/elderly/disabled . . .'. This is followed by the substance of the essay, in which specific incidents or experiences of caring are related, before concluding with a brief reflection on what the young person has learned from the experience, their feelings, or sense of being somehow rewarded or special as a result. It is this final section which reveals most about the students' responses to caring, which must be seen within the context of discourses which help them to make sense of their feelings, but also restrict the extent to which they are able to admit to their less 'acceptable' emotions.

Despite the common discourse, the young people's stories do not all tell exactly the same story, but illustrate a range of responses to being in a caring role. Charles describes fear of infection (Box 1) – an understandable fear reported in medical literature among children close to someone ill or disabled (Becker *et al.* 1998: 4). For Charles, it seems likely that his grandfather died of AIDS, which may have contributed to his fears as well as his shock over the death of his grandfather when apparently so young (Box 1). Anger or resentment may be responses to being chosen to care. For example, Loveless says 'I was feel angry because I was one who was help him' (Box 2). Patikayi writes, 'I used to work like a donkey', which may also reflect resentment at having to do extra farm chores during his uncle's illness, or may simply be a graphic description of how hard he worked.

Many of the young people expressed how much the experience of being closely involved in caring made them want to help more, or to remove the care recipients' suffering. Loveless (Box 2) expresses frustration and helplessness at being unable to stop the illness, loss and pain. He writes, 'I was feel to find some doctor and nases to help him . . . I feel to find some madisan to care of a disease like BP [blood pressure]. Becoase he was suffering with that disease'. Patikayi, similarly frustrated with his inability to help, reports that 'I myself I am petty [pity] for all the people who need to be helped but I can not give them enough help for I am controled by the parents.' It is part of the frustration of youth that despite taking on very adult responsibilities of caring, these young people feel strongly the constraints of their status as minors. Stella, in particular, elaborates her vision and wish to help others (Box 5). This may lead us to speculate that the close encounter with suffering during teenage years which young carers experience may make them more likely than their peers to be motivated towards vocational caring careers.

Other negative responses to caring evident in the young people's caring stories include sadness. Reflecting on caring for his grandfather when he was unable to do even the simplest tasks for himself, Mugowe (Box 4) says, 'I feel sorry about it because it is not a good thing to see other in occasions like that'. It is clearly upsetting and distressing for young people to be close to very sick individuals providing intimate care. Some young people suggest that caring carries a nega-tive stigma and even ridicule. Norman, who cares for an elderly neighbour, explains, 'When I helping him I hate somebody talk to me bad things saying how did you do that and what are your doing.' Given that none of the students suggests there should be stigma or shame attached to the role, it is difficult to assess the extent to which young carers may in fact be (or genuinely feel) ostracised, or whether this, again, reflects wider Christian discourses of doing what is right, regardless of what people think. Students such as Stella (Box 5) construct themselves as almost saintly in their determination not to be deterred by others' mockery.

In return for these saintly acts, being a young carer, as has already been suggested, is understood to bring blessings or rewards. The young person may feel proud and appreciated for what they have done to help someone. Norman says, 'In my life when I am helping him, I can feel happy because he is our grandfather . . . Sometimes when I thinking of him I can fill happy.' Netsai reports how her mother was grateful for the care received and that it has made them closer emotionally: 'When she gets well she gave me many thanks and on top of that she loves me more than other children . . .'.

Emotional closeness and affection between carer and care recipient are evident in other stories. Nancy reports that while looking after her grandmother, 'every lunch I went at that hospital and give her delicious food because I knew her favourate food. Sometimes she refused to eat that food but I speak to her in a polite way until love that food.' The relationships between young people and the individual they are caring for may encompass mutual dependency and benefit. This is acknowledged by other researchers on young carers (Becker *et al.* 1998: 109).

One interesting issue which has been explored in relation to young carers in the North is the degree to which participating in caring contributes to blurring boundaries between childhood and adulthood. It has been suggested that caring means young people shoulder adult responsibilities, and may become 'their parent's parent' (Aldridge and Becker 1995: 120). Caring may thereby make a young person more mature. Among the Zimbabwean students, Stella exhibits maturity and is the most reflective on the impacts of her experience of caring for her aunt (Box 5).

It should be pointed out, however, that Zimbabwean young people, compared with their Northern counterparts, generally have much greater responsibilities in terms of contributing to the physical welfare of their households. Thus, respon-sibility for performing tasks of daily living and household reproduction does not correspond with the adult/child boundary in the Zimbabwean context. However, decision-making is not seen as the corollary of such responsibilities, and most Zimbabwean young people accept adult household members' control over their activities beyond an age at which such influence would be considered appropriate in the North.[9] Despite the portrayal of care recipients as passive victims, and, by extension, the young carer as agent (see Box 1, in which Charles constructs himself as almost the only actor involved in caring for his grand-father), it should not be construed that caring for household members necessarily accords a young person power within a household. Nor is it appropriate to see a young Zimbabwean carer as an 'oxymoron', the caring role necessarily conflicting with normative notions of childhood, in the way that young carers are often portrayed in the North (Stables and Smith forthcoming).

Nonetheless, young people who care are faced very powerfully by issues of life, death, pain and suffering, with for many of them further burdens of grief following bereavement. The full costs of caring for both the young carer (e.g. loss of education, leisure, play, social life, disruption of normal activities and development) and the household/extended family (e.g. loss of income, emotional trauma) are not explored further here, as the students' accounts say relatively little about these aspects of caring and the theme is explored else-where (Robson 1998).

Conclusions

The stories of the young people of Ruchera Secondary School analysed in this chapter suggest that among them girls are more likely than boys to care. They care for short periods for relatives, especially grandparents who are ill or frail, providing all kinds of care. In most cases it is unclear whether the youngsters are lone carers in the reported situations (less likely, but possible), or assisting others to care (more often likely). Boys do very similar types of caring to girls, except that boys do farm work which girls do not report, thus reflecting prevailing gender divisions of labour. Some of the negative responses to caring related by young people include fear, ignorance, anger, resentment, shock, sadness, frustration, helplessness and distress. Pride, appreciation and being loved are among their positive responses.

The research presents the question of whether it is meaningful or useful to define the Zimbabwean young people describing their experiences of caring as 'young carers'. Definitions developed in the North suggest that to be a young carer a young person must be providing a substantial amount of caring on a routine basis to an extent which is restrictive and impacts their childhood nega-tively (Becker et al. 1998: xi). None of the young people writing stories at Ruchera Secondary School had (by virtue of the research design) dropped out of school in order to care. Some of them, like Taurai (Box 3), report caring work that differs little from the routine work generally expected of young people. Others, however, indicate restrictions that caring places on them, for example Charles talks of sacrificing (Box 1), Loveless of staying with his grandfather in

190

hospital for nine months (Box 2), Patikayi of working like a donkey and Netsai of staying at home for two weeks caring for her mother. Having to provide intimate care, experiencing emotional distress, ridicule and misunderstanding from others might be deemed negative impacts on childhood. Thus, we can conclude that some of the young people at least had, or were, undergoing caring experiences that come within the definition of young caring in the (Northern) literature.

However, to draw out these impacts on the young people from their stories and to neglect the wider discourses within which they write would be a mistake. To impose on these Zimbabwean students the definition of 'young carer' is perhaps to pathologise the activities in which they engage. That is not to say that these young people see caring as a part of everyday life. In being asked to write about caring specifically, the students then did so in a way which emphasised that this was an activity separate from the ordinary. But they did not generally imply that caring was a problem in a way that some of the Northern literature, especially by the media (Stables and Smith forthcoming) suggests, in spite of references to the ridicule it may bring. Instead, their own discourse emphasised that to care is to be 'special', to do something which is good, and which can bring rewards.

No claims are made for the material presented here being a fully representative survey, but we believe these stories do provide valuable insights into the caring experiences of young rural Zimbabweans. However, we readily acknowledge that these stories represent little more than just 'the eyes on the hippo'. It is notoriously difficult to find or survey young carers (Becker et al. 1998: 15), and the strategy reported here is just one of a number in the larger project in Zimbabwe researching young carers and the impacts of caring on their lives. It leads us towards recognising both commonalities and differences between the experiences of young people providing home health care in both urban and rural settings in Zimbabwe, as well as between young carers in the North and in the South, which are worth researching further.

It is our hope that the initial work related here will lead others to investigate the situations of young carers elsewhere in the South, particularly pursuing the specific questions raised in this chapter.

Acknowledgements

Elsbeth Robson acknowledges financial support of HSBC Holdings for a research grant awarded by the Royal Geographical Society (with the Institute of British Geographers), while Nicola Ansell acknowledges award of a Wingate scholarship and funds from the Dudley Stamp memorial fund that enabled the research to be conducted. Both authors are grateful for the support of the Geography

Department, University of Zimbabwe while carrying out fieldwork and to Agnes Chidanyika who helped allocate Shona pseudonyms.

Notes

1   In using the labels North and South we remain aware of the shortcomings of a binary distinction which obscures the heterogeneity within each region.
2   Pseudonyms are used throughout for names of the school, persons and local places in order to protect individuals' anonymity.
3   We are aware of debates about ethics and methodologies of researching with children which stress the importance of young people's choice to opt out of and into research without pressure or stigma (Greig and Taylor 1999). However, in the research reported here, the researcher adopted the role of 'teacher', as it was more or less the only possible role in the context of a traditional and conservative school environment. Research attempting to use more innovative methodologies, such as asking students to design and carry out their own research projects, met with far greater resistance from the students.
4   This approach has some similarities with Halldén's (1994) approach of using concepts from literary theory to analyse girls' written stories about their future family life.
5   We have deliberately reproduced the spelling, grammar and syntax of the young people's own writing and provided explanation only where absolutely necessary, in order to retain their genuine accounts. This may make the stories difficult to read, but does preserve and accurately convey their own words.
6   It is perhaps worth questioning the extent to which young Zimbabweans see 'caring' as distinct from everyday domestic work. Though the composition exercise required them to think about giving care to a specific group of people, these young people most probably engage in giving 'care' to fit and healthy members of their families, young and old, on a daily basis. To compartmentalise a group of people on the basis of age, health and ability may be to contribute to a pathologising of age, ill-health and disability which 'traditional' Shona discourses do not support.
7   Alternatively, this may reflect the method employed in collecting these 'stories', eliciting a structure that conforms to a storyline, with a distinct ending.
8   Alternatively, this may be reading meaning into simple errors of grammar – a danger with a 'face value' reading of the stories.
9   For example, many teenagers at the school expressed discomfort at the thought that the law which sets the age of majority at 18 would allow them to act independently of their parents' wishes.

References

Aldridge, J. and Becker, S. (1993) 'Punishing children for caring: the hidden cost of young carers', *Children and Society* 7, 4: 376–87.
——— (1995) 'The rights and wrongs of children who care', in B. Franklin (ed.) *The Handbook of Children's Rights: Comparative Policy and Practice*, London: Routledge.
Ansell, N. (1998) '(Re)constructing gendered identities in the Southern African secondary school', paper presented at the RGS–IBG Conference, University of Surrey, January.

Becker, S., Aldridge, J. and Dearden, C. (1998) *Young Carers and their Families*, Oxford: Blackwell.

Brown, H. and Smith, H. (1993) 'Women caring for people: the mismatch between rhetoric and women's reality?', *Policy and Politics* 2, 13: 185–93.

Foster, G., Makufa, C., Drew, R. and Kralovec, E. (1997) 'Factors leading to the establishment of child-headed households', *Health Transitions Review supplement*.

Frank, J., Tatum, C. and Tucker, S. (1999) *On Small Shoulders: Learning from the Experiences of Former Young Carers*, London/Milton Keynes: Children's Society/Open University.

Gelfand, M. (1979) *Growing Up in Shona Society: From Birth to Marriage*, Gweru, Zimbabwe: Mambo Press.

Glazer, N.Y. (1990) 'The home as workshop: women as amateur nurses and medical care providers', *Gender and Society* 4, 4: 479–99.

Gregor, F. (1997) 'From women to women: nurses, informal caregivers and the gender dimension of health care reform in Canada', *Health and Social Care in the Community* 5, 1: 30–6.

Greig, A. and Taylor, J. (1999) *Doing Research with Children*, London: Sage.

Gwaunza, E., Nyandiya-Bundy, S., Nyangurur, A., Goromonzi, W. and Kamuruko, S. (1994) *A Situation Analysis of Children in Especially Difficult Circumstances in Zimbabwe*, Harare: Zimbabwe Government (Ministry of Public Service, Labour and Social Welfare) with UNICEF assistance.

Halldén, G. (1994) 'Establishing order. Small girls write about family life', *Gender and Education* 6, 1: 3–17.

International Monetary Fund (February 1999) *International Financial Statistics*, Washington: International Monetary Fund.

Muyinda, H., Seeley, J., Pickering, H. and Barton, T. (1997) 'Social aspects of AIDS-related stigma in rural Uganda', *Health and Place* 3, 3: 143–7.

Renfrew, A. (1996) *ESAP and Health: The Effects of the Economic Structural Adjustment Programme on the Health of the People of Zimbabwe*, Harare: Mambo Press in association with Silveira House.

Reynolds, P. (1991) *Dance Civet Cat: Child Labour in the Zambezi Valley*, Athens, Ohio: Ohio University Press; Harare: Baobab Books; London: Zed.

Robson, E. (1998) 'Invisible carers: young people in Zimbabwe's home-based health care', paper presented at the NSF Workshop on Young People's Geographies, 11–15 November, San Diego.

Segal, J. and Simkins, J. (1993) *My Mum Needs Me: Helping Children with Ill or Disabled Parents*, London: Penguin.

Stables, J. and Smith, F. (forthcoming) '"Caught in the Cinderella trap": narratives of disabled parents and young carers', in R. Butler and H. Parr (eds) *Geographies of Disability*, London: Routledge.

UNAIDS (1998) 'Africa: HIV/AIDS Update', online posting <apic@igc.apc.org>, 'AIDS epidemic update December 1998', Washington: Africa Policy Information Centre.

World Bank (1999) *World Development Report 1998/99*, Oxford: Oxford University Press.

11

HOME SWEET HOME?

Street children's sites of belonging

Harriott Beazley

Introduction

At times of estrangement or alienation . . . home is no longer just one
place. It is locations. Home is that place which enables and promotes
varied and ever changing perspectives, a place where one discovers new
ways of seeing reality, frontiers of difference.

(hooks 1991: 149)

In this chapter I focus on children who are living and working on Indonesia's
streets, and their various perceptions of 'home'. By drawing on Scott (1990),
Sibley (1995a, 1995b), and Massey (1992), I discuss how street children experience a variety of emotions about home, how their concepts of that physical
place are tied to their notions of identity, the reasons which often trigger their
desire to return there, and why, almost invariably, they decide to leave again.
Within this context I wish to complicate and question our understanding of the
meaning of 'home'. In particular, I challenge the crucial assumptions embedded
in the 'home ideal' by examining the conflicts which exist between the 'hidden
transcript' of the street kid subculture, and the 'public transcript' of the family
home (see Veness 1992: 445; Scott, 1990).

Street children in Indonesia

I spent fourteen months researching the lives of children who live and work on
the streets of Yogyakarta, a city in central Java.[1] The majority of children visible
on the streets are boys, between the ages of 7 and 17. Girls are also present,
although they are not so readily apparent or numerous as the boys.[2] In Indonesia
street children can be understood partly as an outcome of the country's economic
growth strategy aimed at integrating Indonesia into the global economy. Such

an approach caused Indonesia to experience radical social change, a widening gap between rich and poor, rapid urbanisation, and the marginalisation of millions excluded from the development process. It is in this climate that many children have drifted onto the streets in order to find alternative channels of income.

Financial hardship, however, is not the only reason children start living on the streets. Quite often violence and physical abuse at home forces a child to flee permanently, and some of the reasons given by children for leaving home included being unloved and beaten; alcoholic fathers; pressure to do well at school; absent or separated parents; hostile step parents; the influence of friends and the attraction of street children's subcultures (Beazley 1999). Once on the street, boys earn their money in various ways in different parts of the city, by shoe-shining, busking, scavenging for goods to recycle, or begging.

As in most countries, street children in Indonesia are both spatially and socially excluded by state and society, and portrayed as a 'problem' which needs to be solved. They are also marginalised by the negative perceptions held by mainstream society, who view homeless street children as socially inferior. This is because street children are seen to be abandoned by their families, and to have lost their kinship ties, which are the basis for locating people within Javanese society (Ertanto 1993). A person without such affiliations is considered to be at the lowest possible position in society. As a result, children who have lost all family connections are often considered to be social pariahs infesting the city streets. Their very presence is a challenge to the state's development philosophy, and the ideological construction of the ideal family, home, and child, which it uses for social control (Beazley 1999).

Élite groups in Indonesia have endeavoured to provide models of personal relations and family life which are expected to be in harmony with the nature of 'modern' Indonesian society. Through this 'public transcript' for appropriate behaviour, the home is positioned in Indonesian state discourse as having a fixed agenda of values and interests which are heavily moralised (Scott 1990). This is because the home is the site of child-rearing as well as the central site of consumption, and thus of fundamental importance to the economy and development. The Indonesian state also manages to control the family through various institutions such as the Family Welfare Programme which was established in 1972. The nation can thus be seen to be 'invading the home' by providing cues for behaviour in families, as they relate to the domestic environment (Sibley 1995a: 90).

The dominant concept of family and home as a political agency of control is believed to be threatened by the existence of alternatives such as street children, who constitute a disruption of national space with their 'alien values' (Sibley 1995a: 42). In addition, street children are repeatedly labelled by society and the media as 'neglected' or 'abandoned' by their families, thus placing the blame for their predicament onto their parents, and deflecting blame away from structural or economic explanations. Parents are also labelled as 'lazy' for forcing their children to work on the streets. In this way, street children represent a challenge to the state's enforcement of moral boundaries, as they exist outside of the 'home', and are thus considered to be 'out of place' (Cresswell 1996). They are also highly mobile, and constantly move from place to place, city to city, via Java's extensive railway network.[3] Mobility disturbs the whole notion of order in the eyes of the Indonesian state, which attempts to control movement through its surveillance systems (Beazley 1999).

Such a transgression of the public transcript is considered to be abhorrent and the state subsequently attempts to stigmatise, oppress and conceal street children through a number of social and spatial exclusionary processes. Children who live and work on the streets face the daily threat of violence and abuse at the hands of the police, who regularly clear them off the streets and traffic intersections where they work. In recent years there have been frequent coercive 'cleansing operations' to remove beggars, vagrants and street vendors from the streets, and children are often caught in such crusades. Police and security guards are responsible for confiscating and destroying street children's possessions and merchandise (clothes, shoes, musical instruments and goods to sell), and for verbal abuse, brutal beatings, shaving heads, torture, rape, electric shocks and other abuses which street children frequently report they receive while in custody (Beazley 1999).

In an attempt to find solidarity in the face of this oppression, homeless street children have created their own distinctive social worlds: urban subcultures within Indonesian society, with their own 'hidden transcript': a system of values, beliefs, hierarchies and language (Scott 1990). The creation and maintenance of these street kid subcultures can be seen not as a problem, but as a solution to the variety of problems children face in a world which is hostile to their very existence. As Hebdige (1979: 81) asserts: 'Each subcultural "instance" represents a "solution" to particular problems and contradictions.'

In Yogyakarta, street boys have a name for their own subculture and surrogate family: *Tikyan*, from the Javanese *sithik ning lumayan*, meaning 'a little but enough'.[4] Through belonging to the *Tikyan* subculture, street children are able to create positive self-identities for themselves and to escape feelings of being *malu* (ashamed) about who they are.

Maps of 'home'

My research activities while in Yogyakarta included the collection of 'mental maps' of the city drawn by street children (Beazley 2000, in press). However, despite my requests for them to draw the city in which they lived and worked, some children only wanted to draw maps of their home towns or villages. It seemed that these were places with which they identified more readily than the city in which they lived – even though some of them had been living in Yogyakarta for a long time, sometimes years. During informal interviews and focus group discussions the children talked to me about their feelings for home, and of missing their mothers and younger siblings. It seemed to me that they experienced an array of emotions about 'home' (pleasant memories, warmth, security, fear, anxiety, aversion, excitement, desire) which were all connected to their personal experiences of that physical space (Sibley 1995b).

I recognised the children's desire to draw maps of home as a form of 'regressive nostalgia', as by recreating home in their imaginations and in their maps, they were remembering what it was like to live there (Bammer 1992: x). It was comforting for them and gave them a reassuring 'sense of belonging' (Bammer 1992; Matthews 1992). It was also a way for them to establish their individual identities, as being connected to a particular place. Street children's lives are transient, and their identities are fluid and fractured, depending on where they are and who they are with. By reviving images of home, the boys were creating a place of imagined stability and security: a site of coherence in their hectic lives (Massey 1992). I saw the maps as a way for the children to establish their bearings, to salvage a coherent and centred identity, and to reassert control over what had been lost.

Each child discussed their maps in detail and took time to explain everything to me; where they lived and went to school; where their relatives lived; where they had worked on the street; and where their friends hung out. The maps acted as a catalyst and enabled me to learn more about their personal stories and how they had lived before they left home. The activity also started me wondering what this meant in respect to street children's identities. What were their feelings about home? Did they miss their families? Did they want to return there?

Relph (1976: 6) informs us that to combat feelings of 'placelessness', people have needs for associations with significant places and, at the deepest level, there is always a subconscious association with place. Places are important in the construction of individual identity and are sources of security and identity for both individuals and communities. For street children this identity construction begins with which town or region they are from, through to the places they then occupy in the city. This is because even though they are highly nomadic they still identify with particular places, and construct their individual and communal identities in terms of where they are from (their 'home'), and where they work and hang out in Yogyakarta.

Even though many street children have made the decision to leave their family home as a solution to a personal predicament, their sense of identity with that place continues to inform their own subjectivity and is part of their lived experience. For example, when a child first arrives on the street, the other children always ask where he is from, and he is often given a nickname related to his place of origin: Suvil from Surabaya is known as Suvil Surabaya; Budi from Bandung is Budi Bandung; and Rikki from Blora is simply called Blora.[5] The children also form smaller groups within the *Tikyan* subculture which are related to their places of origin. Further, once living and working in Yogyakarta, street children identify with different territories in the city and call themselves after the places where they work and sleep. For example, the children who work on the main street, Malioboro, are known as *Anak Malioboro* (the Malioboro kids); the children who live and work in the bus terminal are *Anak Terminal*, and the children who live under the Surgawong bridge call themselves *Anak Surgawong*.

Street children regard the places where they live and hang out in the city as safe spaces. They are places in which they can identify with one another and assert their differences and marginality from the mainstream, while creating a sense of belonging to the *Tikyan* subculture (Beazley 1999). This is despite the constant threat of police operations and physical oppression within these spaces, because the security they feel is an 'emotional and psychological safety that comes from being in an area which has some sense of belonging or social control, even in the occasional absence of physical control' (Myslik 1996: 168). By creating these feelings of security and belonging, such spaces have, in a sense, become their alternative homes.

Two realities

'When did you last see your mother?' someone asked me. Someone who was walking with me in the city. I didn't want to tell her; I thought in the city a past was precisely that. Past . . . 'Don't you ever think about going back?' Silly question. I'm always thinking of going back . . . People do go back, but they don't survive, because two realities are claiming them at the same time . . . Going back after a long time will make you mad, because the people that you left behind do not like to think of you changed, will treat you as they always did, accuse you of being indifferent when you are only different.

(Winterson 1985: 160–1)

The majority of street kids start their lives on the street as 'ordinary' mainstream children, and slowly become socialised and imbued in the street life and *Tikyan* subculture. Over time their identities shift and accord with the expectations of their own reference group, as they become socialised to alternative values and ways of existing. Such values include wearing earrings and tattoos, smoking, swearing, having a dirty appearance and practising 'free sex' (Beazley 1999). Consequently, the behaviour patterns of their subculture often represent a challenge to the dominant culture's rules and values, although the children are not completely isolated from the dominant culture. This is because they still interact with the external world of mainstream society, which socialises them with the appropriate ways to behave, and reminds them of a cultural 'alternative' to their own existence. As a result, two sets of standards are presented to the children, which can create a clash of values within them, as 'two realities are claiming them at the same time': their subculture and the dominant culture.

Through various people on the street, Non-Governmental Organisation (NGO) workers, and the mainstream cultural apparatus (films, radio, TV, video games, and a diverse mass media), street kids are instructed in dominant cultural norms, with the attendant prescriptions of the 'proper' behaviour for young men and women. This constant contact with external socialising influences creates a kind of parallel socialisation which Hannerz (1969: 137) calls 'biculturation'. Biculturation happens when mainstream definitions are stated with such authority that they have a pervasive impact on subcultural groups, who sometimes want to participate in the mainstream consumer culture and conform to the 'alternative' world which most of them once knew. In short, they sometimes yearn for stability and wish to be 'normal'.

This is particularly true as street kids reach the 'liminal period' of adolescence, and are increasingly alienated from society (James 1986). As they get older, and their bodies transform into those of young men, they no longer have the 'cute' appearance of a young child. Instead, they begin to resemble teenage 'thugs', and are often regarded with distrust by society. This makes it considerably more difficult to earn an income on the street. Thus, as they reach adolescence, street youth find it harder to earn money, and due to their appearance they are increasingly alienated by society and the state apparatus (police and security guards) who perceive them as thugs or hoodlums. For these reasons the street becomes increasingly less attractive and more dangerous for street youth, who simultaneously become disoriented and unsure about themselves and their individual identities. They start to see their life as dominant society sees it, and consider returning to the mainstream. One boy, Arek (age 15), for example, told me that when he was younger he did not care about what people thought of him sleeping on the streets, but now when he wakes up late in the morning and sees people staring at him, he feels ashamed.

It is at these perplexing times that street children will often consider going home and going back to school or getting a regular job. On the street their lives are so unpredictable, and in a constant state of flux. They often seek a centred and coherent identity and need to remember home to remember who they are, as 'home serves as a boundary of the self' (Sibley 1995a: 94). While on the street, 'home' is an image which they have mentally constructed as a comfort zone and as a form of refuge against the outside world. In particular, I often heard the children say that they missed their mothers. It seemed that she was the personification of that place which (in their mind's-eye) had not changed (Massey 1992).

Missing home

Sometimes an event or a chance meeting will trigger memories and familiar feelings of home. It may be from watching a film or from forming a close relationship with a mother figure. For example, some children told me that I reminded them of their mothers, aunts or elder sisters. Often, if I saw a child looking perplexed or upset and asked them what the matter was, the answer would be 'I'm thinking about going home', or 'I'm thinking about mum'. They also said that they missed their younger brothers and sisters. Street children's attachment to an NGO can also trigger fond memories and remind them of aspects of living at home which they miss. In a way NGOs dealing with street kids are like surrogate families, which can cause some children to start thinking about their own families. Heri, for example, went home for the first time in years as a direct result of visiting the street kid organisation, *Girli*, in Yogyakarta. He said that the care and attention the workers gave the children reminded him of his mother and home.

In Yogyakarta the desire to conform to the mainstream and to go home was particularly strong during the period of *Puasa* (or *Rahmadan*): the Muslim fasting month. The children get caught up in the festive fervour, which is evidence that external influences of mainstream cultural traditions do make an impression on them, and that they do retain memories about that way of life. Many children attempt to fast at this time, although it often proves to be impossible, as they have to work in the hot sun and need to drink.

It is also during *Puasa* that dozens of children contemplate going home to ask forgiveness (*minta ma'af*) for their 'sins', which is a tradition during *Lebaran*. *Lebaran* is the religious festival that marks the end of the fasting month and is the time when, once a year, people are expected to go home to their villages and spend time with their family. It is a very special time to be able to ask forgiveness from one's parents and siblings for all one's mistakes over the past year, and to bring home money and gifts to the family in the village (Ertanto 1996; Sari 1997). Yogyakarta is practically deserted at this time, as thousands of people pour into the countryside to visit their home towns and villages. If they cannot go home, however, street kids will often go to Jakarta where the income-earning possibilities are better during the holiday season. Others 'hide' until after the holiday period. The kids told me that this was because they were ashamed to be seen alone, as it is a sign that they have nowhere to go and that they are unwanted.

Home, therefore, is an imagined point of orientation, a centre of safety and security, 'a stable physical centre of one's universe – a safe place to leave and to return to' (Relph 1976: 83; Rapport 1995). Sibley (1995a: 92–4), however, warns against such a 'cosy' and idealistic view of home and observes that, due to power relations within the home, it can actually be 'another space of exclusion'. Similarly for street kids who try to return, the imagined stability and security of home often proves to be false.

Problems with going home

There are street children who are able go home and visit their mothers, and who do so fairly regularly. Others, however, cannot, because although they may desperately want to, they face a number of dilemmas before they even get there. Some, for example, do not have any parents or do not know where they are, either because it has been so long since they left, and they cannot remember, or because their parents have moved without a forwarding address. This was sometimes the case with street children from poor families whose *kampung* (poor urban housing) had been evicted or razed during government-sponsored 'improvement' programmes. Other children's parents survive in the informal economy and their lives are as transient as their children's. Alec (age 12), for example, wanted to go home to Surabaya at *Lebaran*, but his father was in prison and he did not know where his mother lived. When he eventually found her, she was living and working in a brothel. His sister had been given away, as his mother could not afford to keep her.

Still others do not want to go home because they were abandoned by their mothers or they have been rejected by their parents when they have tried to return in the past. Cecak's (age 16) mother left him when he was small. She remarried and had more children with her new husband – this is something about which he feels very bitter. He explained that he wants to kill mothers who have not protected their children and that he hates all mothers, seeing them as satanic beings who abandon their children. A number of children are also afraid to go home because they fear rejection or retribution for something they did wrong in the past. Children I knew had stolen money, jewellery or a bicycle to sell in order to get the money to leave, and were adamant that they could not return, as they would not be forgiven for such a misdeed. Abusive stepfathers or mothers were also a dominant reason for a child's fear of home and, all too frequently, children told stories of being beaten up by their (step) parents, uncles or elder brothers when they tried to return.

Children often recounted tales of confronting angry parents when they went home. During *Lebaran* in 1997, Rian (age 12) went back to his parents in Malang. Even though they had not seen him for over a year, the first thing his stepmother said to him was: 'Where's the money? I don't see you for a year, and then you come back empty handed.' Danang (age 12) had a similar story: 'I was immediately shouted at when I got home, because of the problem of money'. Little Budi (age 11) was very upset one day when he went home and his stepmother told him to leave again. He came back to Yogyakarta and at the *Girli* training house he drew a *batik* picture of the event (Plate 11.1).

So, although street children sometimes fantasise about returning home, once they get there they often find that it is another space from which they are excluded. This is particularly true if they come home looking poor and dishev- elled. Frequently, children would complain to me that they could not go home because their clothes were filthy, or because they were wearing the same clothes they had left in a year before. One child explained to me: 'There's a culture of shame if you go home without having succeeded'. The boys also told me that it was not only their own personal pride which stopped them from returning to their *kampung*, but that their parents would be furious if they returned empty- handed. Far from being greeted as the 'prodigal son', a street boy will often be punished if he is dressed like a vagrant because he is shaming the family's repu- tation. There is a Javanese saying: 'A father is responsible for his child's behaviour' and, in Indonesian society, the father is often held to blame for a street child's circumstances (Ertanto 1996).

Plate 11.1 Budi's (age 11) batik drawing of being evicted by his stepmother: The caption reads 'Budi is evicted by his mother' ('Budi diusir ibunya'), with his step- mother shouting 'GO!' ('Pergi'). © Harriott Beazley

Homeward bound

Within this context, in Javanese culture it is quite usual to take home with you an impressive friend to show that one's connections are good (Ertanto 1996). It was partly for these reasons that a number of boys asked me to go with them: as a white foreigner (*bule*) I was something to take home, which the children knew would give them prestige in their village. I also learnt that I was a form of protection against their violent parents, and children will often take a friend or NGO worker with them for this reason.

I made a number of trips with boys to their families in Sumatra, central Java and the outskirts of Jakarta. All of these children, except one, came from very poor families. When they returned to their *kampung*, however, they judged their homes with their new city eyes. They all told me that their home lives were dull, backward or boring. When Topo (age 13) took me to his riverside abode, for example, he said: 'My house is crap isn't it Hattie?' Wiwit (age 15) (who had not been home for four years) was really shocked at how 'backward' his *kampung* in south Sumatra was, and how poor his family were. He viewed his city as a 'failure'. On our way back to Jakarta he drew a picture and wrote:

> My city is a place that has been
> failed by Indonesian society and
> I like Jakarta and
> I also like living in
> the street.

Other children were upset by the conditions in which their families lived. Arief told me that he did not like going home, because he could not tolerate seeing his grandmother and younger sisters so poor and in need. Further, Suvil (age 15), Aat (age 14), Arief (age 16) and Dede (age 12) all said that they did not want to work in the rice paddies, and that life at home was too hard and boring. When I went home with Wiwit, his brother offered him a job as a passenger recruiter on the buses, but he turned it down, saying he would not earn enough money, and that he could earn more on the street busking. Most of the boys told me that in spite of the dangers and harsh life on the street, they were happier living there than at home. In fact, many claimed to be better off on the street than they or their siblings were at home. Living at home usually means having no money to spend, feeling out of place and isolated, facing problems of adaptation and membership with one's siblings, being beaten or verbally abused, facing poverty and horrid food, missing one's friends (and feeling alienated and 'different' from old ones), and having to witness one's mother and siblings suffering in poverty. After a visit home with Suvil (age 15) I asked him if he would ever live at home again. 'Never' he replied, 'why would I want to live like that?' 'Like what?' I asked. 'All they do is sleep and work' he said.

Sorio (age 11), however, was from a rich family. When I went with him to visit his family home in a salubrious suburb of Jakarta I was amazed at the luxury he had left behind for a life on the streets. He had an air-conditioned house, computer games, servants and a chauffeur-driven Toyota Landcruiser to drive him around. His mother and grandmother told me that they could not stop him from running away a year before.

Most of the relatives I met told me of how upset they had been when their son or sibling had first *kabur* (ran away). They searched through the markets, went to the police, asked around, and went to the hospital if an unknown child died. They told me that the authorities had been very unhelpful, and were not interested in missing children. This is very unlike England or Australia, where if an 11-year-old boy goes missing it is reported in the national newspapers daily. In Indonesia it appears that the only time the police will pay any attention to children living and working on the streets is when they are 'cleaning them up', in order to tidy the image of the city. Even when they take them into custody, they do not attempt to find out where they are from, and usually just hold them for a night or two, often abusing them, and then telling them to 'go home'.

Some children do not want to be found, and change their names or hide from their parents if they know they are looking for them. Sorio (age 11) for example, said he was happier living on the street, and wrote a letter to his grandmother, begging her to leave him alone:

> For Grandma, who I love: . . . Grandma, there's no point being worried about Sorio's situation in Yogya. Sorio is very happy here and Grandma needn't bother coming here because Sorio is going to school in Yogya and Grandma, don't hassle Sorio in Yogya. If Grandma really loves Sorio then Grandma won't take the trouble to come to Yogya, because Sorio doesn't want to be bothered. If Grandma wants to send Sorio some snack money, you can send it in an envelope to *Girli* [the street kid NGO] . . . From your grandson in Yogyakarta. Don't say anything to Mama or Father. Please!!

Sorio told me he was bored at home and that his father beat him. He had made the decision to leave when he saw his father beating his pregnant mother. He told me that he much preferred living on the street, as he could be independent and with his friends. For these reasons his periodic visits home never lasted long and became progressively less frequent. As was the case for many street kids, the allure of the street and the *Tikyan* subculture were far stronger for Sorio than the attraction of the family home.

When they went on a visit home, however, all of the children I accompanied attempted to conform to the performance which was expected of them by their families. They were all very careful with their appearance, hid their tattoos, took out their earrings and put on clean clothes, and were anxious to look clean and presentable and to take back gifts. By wearing earrings and tattoos on the street, the children are conforming to the *Tikyan* style and are communicating a gesture of defiance to the outside (and anonymous) world (Beazley 1999). They do not necessarily wish to present this same defiance to their own families, or else they dare not. Some children even lied to their parents (or distorted the truth) about their circumstances. Wiwit, for example, told his family that he was at school in Yogyakarta. Heri told his mother that he was a road sweeper, as he was ashamed to say that he was still a street child after seven years (which was the last time he went home).

In this respect, one thing which particularly struck me were the changing identities and personalities of the children when they went home. In every case the boys reverted to a 'child-like' identity and became much quieter and deferent in front of their parents.[6] Living on the streets results in changes happening to a child's identity, as he develops specific survival competencies, and adapts to the norms and values of the *Tikyan* subculture. At home the street kids I accompanied discarded their specific street images and, with the air of self-confidence gone, they were very unlike the assertive, loud, know-it-all youngsters I knew on the street.

Rules

> The home is one place where children are subject to controls by parents over the use of space and time and where the child attempts to carve out its own spaces and set its own times. The possibilities of conflict here are considerable. Children may find the domestic regime oppressive because of rigid parental control of space, the availability of space in the home may limit opportunities for children to secure privacy.
>
> (Sibley 1995b: 129)

Sibley (1995b) has noted that problems at home often arise in families where parents have oppressive or alienating relationships with their children. He describes these families as 'positional', which means that power is vested in position in the family. Traditionally, in Indonesia children are expected to be silent and deferent, and a father is distant and cool towards his children and relates to them in an authoritarian manner (Sullivan 1994). This attitude can manifest itself in the imposition of arbitrary rules and in the issuing of instructions and punishment without explanation. Such power relations have a direct impact on how street children experience home when they go there.

For many street kids, one rule is too many. This is because they fiercely value their independence and there is a pervasive ideology of individualism within the street child subculture (Beazley 1999). All street kids seek adventure, autonomy and freedom. This desire increases the longer they have been living on the street, and as their *Tikyan* boundaries are reinforced they shift further away from those which are deemed 'acceptable' in the home. Children often complained to me that there is too much discipline at home and that they cannot fit in with the temporal and spatial restrictions placed upon them. As Mohammed (age 12) once complained to me when he returned to the streets after a visit home: 'There are so many rules at home. There's a time for getting up, a time for eating, a time for washing, a time for praying, a time for school. There is too much discipline.'

In addition to these restrictive boundaries, street children also find that when they go home the survival competencies which they have developed on the street are not recognised or appreciated by their parents or siblings. They are treated as children, as though they have not changed, which strips them of their feelings of status and self-worth. Consequently, they long to return to the street world where they are accepted, and where they have some status within the *Tikyan* group as well as autonomy from their families.

When I said I was going, none of the children with whom I went home wanted to stay, instead they left when I did. A few said that they would have 'certainly been beaten' if they had not left with me. I had been their protection. Twice when I was at a boy's house, the boy would start agitating to leave before his stepfather or father came home from work. This was the case with Topo (age 13) when we went to visit his mother in Jakarta. He was 7 when she remarried and moved to Jakarta. She left Topo on the streets in Yogykarta, as her new husband would not have him living with them because he was 'too stubborn'. After our visit, Topo was very pensive and started to complain about toothache. I asked him what was wrong and he said that it was the same tooth that his stepfather had broken when he had beaten him years before. The next day I met him in the street. The whole left side of his face was swollen, and I thought he had been in a fight. Topo said no, it was just toothache. When I looked into his mouth, I could see what had happened: he had gouged a big hole where the broken tooth had been until it was sore and bleeding. Such self-mutilation was not an uncommon reaction to the experience of going home: a number of other children hurt themselves in different ways, while others had bouts of drinking or excessive glue consumption after visiting home.

Thus, despite their desire for some stability and to be accepted by the mainstream, many street kids are unable to tolerate the strong boundaries enforced by their 'positional' parents, particularly when they have become used to the freedom of the streets and living as they please. When they go home the children discover that the boundaries and values they have developed on the street are not acceptable in the family home. Instead, they have to carry out an 'onstage performance' (no earrings, tattoos, scruffy clothes, smoking, sniffing glue, bad language, taking drugs, etc.): a different identity which is expected of them by their families. It is extraordinarily difficult for street children to maintain this 'public transcript' of the dominant culture for long (Scott 1990: 123). After a while (a few months, weeks, days, or even a few hours) they cannot stand it any more. They long to return to the 'hidden transcript' of the street world, where they do not have to be bothered with the imposition of temporal regulations and spatial boundaries of the home. As Danang (age 12), who has been on the street since he was 7, explained to me:

> If a child has been on the street for only a few months then there is a good chance that he will be able to live at home again. If, however, it has been as long as a year since he left then it will be very hard for him to stay there. He will miss his friends and become bored, and will long to go back to the street life and be free.

Conclusion

In this chapter I have discussed street children's feelings about the family home, and how they are increasingly caught between two irreconcilable and opposed lifestyles: the street and the family home. I have shown how for most street kids the street may not necessarily be the way of life that they want to engage in, but it becomes their preferred option between 'two realities', and eventually becomes their 'home'.

Most children I met had constructed their own identities to belong to a place (a village, a town or a house) which they called 'home' – an image to which they sometimes retreated in moments of loneliness or despondency. As the children's experiences show, however, these idealised conceptions were usually seen through rose-coloured spectacles, which were shattered if they tried to return. This is because home is not always the idyllic place of safety and shelter for which the children yearn, but can be a place of fear and danger which they can only inhabit occasionally, if at all (Bammer 1992). If a street child does go home he may find that his parents are angry with him, that his achievements are not recognised, that he is beaten or chastised for coming back without money, or that he is not even wanted. Or he may just find that life at home is intolerably dull.

Often, street children are fleeing poverty or abusive home situations, and on the street they find others with similar experiences to their own. These companions form an alternative family, which has its own values, emotional support, feelings of membership, and an empathetic understanding that they can no longer find at home. In Yogyakarta, the *Tikyan* community is a stable basis for a new identity and a site of belonging for street boys. It is a means through which alienated children can voice their collective indignation at the way they are treated by the state and dominant society, and from where they can collectively refuse and subvert state ideology. It is also from within the *Tikyan* subculture that street children begin to regard specific shifting locations (where they earn money, form friendships, find enjoyment, and feel safe) as their new 'homes'. In this way 'home becomes more of a concept than a material space' (Arantes 1996: 161).

Due to their feelings of membership in the *Tikyan*, abandonment of the street is difficult, or even impossible for street boys. It has become a central part of their lives, and a way of life with which they are most familiar. Over the months or years, they have made a strong investment in the subculture. It is where they have developed intense emotional bonds, an individual status, a position in the hierarchy and valuable connections for themselves. In mainstream society they have nothing. This compounds their problems of making a transition to the mainstream.[7]

Thus, despite the risks and dangers which they must face, including violence and abuse from state and society, once they have been living on the street for a long time it is very difficult for street children and youth to go home or to re-assimilate into dominant society. Even though they know what is expected of them by the mainstream (and sometimes they wish to conform to those expectations), ultimately they do not feel the moral pressure and social control of the dominant culture as strongly as the obligation to conform to their own subculture (Beazley 1999).

In summary, street kids in Indonesia actively challenge and subvert the state's ideological constructions of the ideal home, family and child. They do this by not only working on the streets, but also by rejecting the family home and creating an alternative family (the *Tikyan*) which inhabits numerous 'homes' (a market, the main street or square, the bus terminal or train station, a traffic intersection, or space under a bridge). Thus, for children living on the streets, home is no longer just one location or physical space in a village or a *kampung*. Instead, it has become a 'moveable concept', attached to the *Tikyan* subculture which the children have created, and which is spread across a variety of locations throughout the city and country (Bammer 1992: ix; hooks 1991). These locations are places which the kids inhabit with others, and which change with the *Tikyan*'s shifting geographies of social relations and identities (Massey, 1992). In effect, certain spaces in the city have become 'home in the public space' (Arantes 1996: 86), where an alienated child is able to survive, and to feel as though he exists in a world that would rather he did not.

Acknowledgements

I would like to thank the children and workers of the NGO Girli in Yogyakarta for assisting me in my research. Special thanks also to Alison Murray, Tim Curtis and Monica Taylor for their comments and suggestions on earlier drafts of this chapter. Of course, any shortcomings in the chapter are solely my responsibility, and in no way should be attributed to anyone else.

Notes

1   The material in this chapter was produced as part of a Ph.D. thesis in Human Geography at the Australian National University. Fieldwork was conducted during the periods June to September 1995, July 1996 to March 1997, and July to August 1997. Funding was provided by the Commonwealth Scholarship and Fellowship Plan (Australia), and the Australian National University.

2   This chapter specifically focuses on the lives of street boys. Elsewhere, I have written about the lives of street girls and how they live and operate in different parts of the city to street boys (Beazley 1998; 2000). One reason street girls are less visible is because they do not engage in the same income-earning activities as the boys. Usually, street girls survive by being looked after by their 'boyfriends': their principle form of income and protection.

3   The children travel vast distances around Java, often stowing away on goods trains, or by sitting on the roof or in between the carriages of passenger trains.

4   A distinctive part of the street child subculture is the use of their own slang or private language. The vocabulary relates to events, activities and objects which are regularly used by the children. The *Tikyan* language creates a realm of autonomy and solidarity, reinforcing a sense of belonging, and excluding outsiders who cannot understand (Beazley 1999).

5   In order to protect their identities, all the street children's names in this chapter are pseudonyms.

6   My choice of the word 'deferent' is deliberate: As Scott (1990: 24) tells us: 'Acts of deference . . . are intended in some sense to convey the outward impression of conformity with standards conveyed with superiors'.

7   See also Hannerz (1969), who observed a similar dilemma for young men caught up in the ghetto culture of Harlem.

References

Arantes, A. (1996) 'The war of places: symbolic boundaries and liminalities in urban space', *Theory, Culture and Society* 13, 4: 81–92.

Bammer, A. (1992) 'Editorial', *New Formations* 17, Summer: vii–xi.

Beazley, H. (1998) 'Subcultures of resistance: street children's conception and use of space in Indonesia', *Malaysian Journal of Tropical Geography* 29, 1, June.

—— (1999) '"A little but enough": street children's subcultures in Yogyakarta, Indonesia', unpublished Ph.D. thesis, Australian National University, Canberra.

—— (2000) 'Street boys in Yogyakarta: social and spatial exclusion in the public spaces of the city', in S. Watson and G. Bridge (eds) *Blackwell Companion to Urban Studies*, London: Blackwell.

Cresswell, T. (1996) *In Place/Out of Place: Geography, Ideology and Transgression*, Minneapolis: University of Minnesota Press.

Ertanto, B. (1993) 'Kere Ki Sah Mati. Yen Mati Ngrepoti: Studi Mengenai Anak Jalanan Dan Perubahan Sosial', paper presented to seminar Posisi Anak-Anak Dalam Konteks Sosial Di Indonesia, Universitas Gadjah Mada, Yogyakarta, 27 November.

—— (1996) *Tekyan Atau Anak Haram Keluarga*, Yogyakarta: Humana.

Hannerz, U. (1969) *Soulside: Inquiries into Ghetto Culture and Community*, New York: Colombia University Press.

Hebdige, D. (1979) *Subculture: The Meaning of Style*, London: Methuen.

hooks, b. (1991) *Yearning: Race, Gender, and Cultural Politics*, Boston: South End Press.

James, A. (1986) 'Learning to belong: the boundaries of adolescence', in A.P. Cohen, (ed.) *Symbolising Boundaries: Identity and Diversity in British Cultures*, Manchester: Manchester University Press.

Massey, D. (1992) 'A place called home', *New Formations* 17, Summer: 3–15.

Matthews. M.H. (1992) *Making Sense of Place: Children's Understanding of Large-Scale Environments*, Hemel Hempstead: Harvester Wheatsheaf.

Myslik, W. (1996) 'Renegotiating the social/sexual identities of places: gay communities as safe havens or cities of resistance?', in N. Duncan (ed.) *Body Space: Destabilizing Geographies*, London: Routledge.

Rapport, N. (1995) 'Migrant selves and stereotypes: personal context in a postmodern world', in S. Pile and N. Thrift (eds) *Mapping the Subject: Geographies of Cultural Transformation*, London: Routledge.

Relph, E. (1976) *Place and Placelessness*, London: Pion.

Sari, M. (1997) 'Pulang Kampung', *Gatra*, No. 16, Tahun III, 8 March (Jakarta).

Scott, J.C. (1990) *Domination and the Arts of Resistance: Hidden Transcripts*, New Haven: Yale University Press.

Sibley, D. (1995a) *Geographies of Exclusion: Society and Difference in the West*, London: Routledge.

—— (1995b) 'Families and domestic routines: constructing the boundaries of childhood', in S. Pile and N. Thrift (eds) *Mapping the Subject: Geographies of Cultural Transformation*, London: Routledge.

Sullivan, N. (1994) *Masters and Managers: A Study of Gender Relations in Urban Java*, St Leonards: Allen and Unwin.

Veness, A.R. (1992) 'Home and homelessness in the United States: changing ideals and realities', *Environment and Planning D: Society and Space* 10: 445–68.

Winterson, J. (1985) *Oranges are Not the Only Fruit*, London: Pandora.

Part 3

LEARNING

12

PLAYING THE PART

Performing gender in America's playgrounds

Elizabeth A. Gagen

Introduction

A central theme in both recent and long-standing work on children and young people is the institutionalisation of major aspects of their daily experience (for example Rivlin and Wolfe 1985; special issue of *Environment and Planning A* in press). From child-care, through formal schooling, to the myriad ways in which children's leisure is organised and contained, children's lives are commonly understood as closely orchestrated by institutional frameworks. In accordance with this assumption, it follows that scholars have been interested in exploring the objectives contained within the agenda of such institutions. With reference to the school curriculum, James, Jenks and Prout (1998: 42) state that they are 'never accidental and certainly not arbitrary', on the contrary, they represent 'social and political structures, containing assumptions about how people (that is largely children) ought best to be'. Educational establishments, loosely defined, represent the spaces through which societies expect children to be socialised toward adult norms. '[I]n both innovative and traditional schools', argues Aitken, 'good citizens are children who conform to social norms and group behavior defined as appropriate by the authority' (Aitken 1994: 89). Learning environments, then, are often the spaces through which children become aware of, and begin reproducing, social identities that circulate through broader social space.

My focus here is the playground as a learning environment scripted by implicit gender norms. Drawing from Judith Butler's (1990) work on gender identity and performativity, I explore the construction and operation of playgrounds in early twentieth-century cities in the United States. My aim is two-fold: first, to explore the particular dimensions of gender ideals that were sought through recreational practices by play leaders; and second, to explore how such ideals were elicited through a spatial regime. I use Butler's theory to explore how heterosexist gender norms are sought through spatial practice, and emerge explicitly in learning environments like playgrounds. While the performative nature of gender identity can be explored at all ages, it is particularly fruitful with regard to children. In spaces where learning is the predominant enterprise, adults' effort to induce particular performances confirms, first, that the aim of such practices is to acculturate children toward social norms, and second, that the alignment of certain gender performances with certain sexualised bodies is not inevitable, nor natural, but closely managed.

Playgrounds and performativity

Rather than assuming that rules and values cultivated through learning environments inevitably succeed in socialising children, scholars are increasingly reorienting research to focus on the ways in which children rework such practices on their own terms. Adult frameworks are often 'creative potential', Aitken suggests, 'as opposed to . . . spaces through which they [children] are oppressed and subjugated' (Aitken 1994: 9). This emphasis is evident in the recent volume, *Cool Places: Geographies of Youth Culture*, which in addition to containing a section on sites of youth resistance, is influenced throughout by a desire to get beyond notions of unidirectional socialisation processes (Ruddick 1996; Valentine 1996, 1997). From this tension I find it useful to turn to performativity. Butler's notion of performative gender is a useful bridge between overly determining institutional or biological (including psychological) explanations.

For Butler (1988, 1990, 1993), gender exists through particular bodily acts that are circumscribed as normal through visible repetition and socially endorsed rewards and punishments. Neither entirely historically given, nor completely individual, gender identity exists as a socially scripted performance of acts that are recognisable as either 'male' or 'female'. The historical nature of norms is not stable, but subject to reinterpretation in any given setting, thus enactment always assumes interpretation (Butler 1988, 1990). While this theory allows for the prescriptive nature of identity, there remains some autonomy on the part of the individual to act out the script in unique ways.

Within the literature on identity, the body has held high rank as the marker of identity. Butler (1990), however, critiques the common, but erroneous, premise, derived from Foucault, that the prediscursively 'sexed' body is the raw material on which culturally constructed signifiers, like gender, are inscribed. Instead she asks: 'To what extent does the body *come into being* in and through the mark(s) of gender?' (Butler 1990: 8, emphasis in original). This is an important amendment, as the body itself only becomes meaningful when enacted as gendered. To successfully enact one's 'proper' gender, then, is to receive license as a legitimate subject. And thus the 'sexing' of bodies is easily overlooked. As Butler states 'sex' is:

not simply what one has, or a static description of what one is: it will be one of the norms by which the 'one' becomes viable at all, that which qualifies a body for life within the domain of cultural intelligibility.

(Butler 1993: 2)

The coincident mapping of acts or performances that are understood as gendered, onto bodies that are naively assumed to be already 'sexed' male or female, sustains conventional gender norms. The regulation of gender identity relies upon the reification of a fictitious biology; thus both 'gender' *and* 'sex' are regulated fictions whose co-dependency secures normative gender/sexual identities.

This system is perpetuated by two connected processes: first, through the actual acts themselves; and second, by the assumption that these acts are generated from within the 'self' and are therefore inevitable. Butler (1988: 523) asks us to 'consider the way in which this gendering of the body occurs. My suggestion', she continues 'is that the body becomes its gender through a series of acts which are renewed, revised, and consolidated through time'. Elsewhere she writes:

The effect of gender is produced through the stylization of the body and, hence, must be understood as the mundane way in which bodily gestures, movements, and styles of various kinds constitute the illusion of an abiding gendered self.

(Butler 1990: 140).

Through this process we come to judge certain acts, pursued by one or other of two body types, as necessarily male or female. The corporeal acts of gender are seen to be coherent – as somehow instinctive to that body – precisely because of the coincident pattern of behaviour associated with one body type. It is this tautology that secures conventional gender politics. If the cause of these performances is believed to be located in the 'self', 'then the political regulations and disciplinary practices which produce that ostensibly coherent gender are effectively displaced from view' (Butler 1990: 136). As such, the apparent naturalness of gender and its attendant performances are left intact.

My interest here is in how strategies and practices that construct learning environments contribute to the maintenance of normative gender performances. If the history of corporeal styles, that is the acts that are already 'socially established', set the limit for gender performances (Butler 1990: 140), then institutions of childhood are the likely spaces through which those available norms are established. In this chapter I explore the ways in which programs of play in early twentieth-century playgrounds evidence a commitment to the constitution of gendered subjects. By spatially segregating girls and boys and directing different activities in each space, girls were coached to perform 'feminine' acts and boys 'masculine'. Together they present a consistent effect of imaginary coherence. In the remainder of this chapter I use the records of playground organisers in Cambridge, Massachusetts, 1902–11, to explore the political dimensions of gender performances that children were taught in playgrounds.

The playground movement and playgrounds in Cambridge, Massachusetts

The widespread institution of public, supervised playgrounds in US cities at the end of the nineteenth century was the result of organised philanthropic effort. Like many organisations that aimed to solve the burgeoning ills of industrialisation and immigration, playground reformers targeted children as the most efficient route to social salvation. The rapid growth of immigrant populations in densely packed urban areas induced widespread moral panic, as reformers plotted crude connections between disease, disorder and morality (Boyer 1978; Sibley 1995). Within countless reform efforts – from temperance and anti-prostitution, to the Immigration Restriction League – children held a unique status. As Ward states, 'Children in particular were regarded as critical targets of reform since they had not fully assimilated the deviant values of their parents' (Ward 1989: 24). Reformers' confidence in the inherent plasticity of children gave rise to multiple institutions, including the reformatory school, through which children could be transformed, and would thus transform, the future of society (see Platt 1977; Ploszajska 1994). Playgrounds, however, had a wider scope. They were not created specifically to target the 'delinquent' child, but offered supervised play spaces for all children who would otherwise play on the street. Clarence Rainwater, a contemporary sociologist of the playground movement, writes that:

> the function of the playgrounds was . . . to keep children off the streets and thus away from certain physical and moral dangers by inducting them into safer places for play, a social situation in which a constructive use of materials and social forces was made for the development of personal and group life.
>
> (Rainwater 1922: 265)

Children were, he writes, to be saved by 'inducting them into activities designed to promote certain behavior' (*ibid.*: 53).

The Playground Association of America (PAA) was founded in 1906, providing centralised organisation to oversee the development of playgrounds across America. This organisation was, in part, responsible for the subsequent surge in the popularity of playgrounds (Cavallo 1981). According to a PAA leaflet in 1909, entitled 'Playground Facts', the number of cities providing playgrounds increased from 90 in 1908, to 336 in 1909.[1] In each city local committees controlled playground production. Initially, local organisations were predominantly philanthropic, later yielding to municipal authorities (Cavallo 1981; Goodman 1979). Following this pattern, in 1902, the Mother's Club of Cambridge (MCC) created a sub-committee – the Playgrounds Committee – to manage the introduction of playgrounds throughout the city. In 1911 they handed control over to the municipal authorities. During that time the Committee established and maintained a total of sixteen playgrounds. All playgrounds were supervised and operated between June and September when the incidence of children playing on the street was thought to be at its peak. The Committee took responsibility for selecting the site, organising the activities, and securing apparatus. In addition, they raised funds to pay the rent for each site, where needed, and supported at least one supervisor per site.

The Committee were already well established by the time the PAA was founded in 1906. They were not, therefore, directly affiliated with the national organisation, but subscribed to the broader ideals of the movement. The Committee echoed the PAA's motives for playground provision as a solution to broader social ills. In calling for some kind of provision for children during the summer months, one member of the MCC argued that playgrounds are needed 'for those children who would otherwise have been hanging about the streets at that time and probably making the most of that mischief which Satan is supposed to find "for idle hands to do"' (Mrs. Almy's report to MCC n.d.).[2] In addition to fearing unsupervised play for the very reason it was unsupervised, the Committee were concerned specifically with immigrant children. An article from *The Record*, 1909, by the husband of one Committee member, states their vested interests in playgrounds: 'We have a very crowded city, an enormous number of boys, either foreign born, or born of foreign parents, and we must give them some chance to play and learn American ways and to keep them off the streets' (*The Record*, 9 February 1909). To maximise their effect, the Committee selected sites in immigrant neighbourhoods (East Cambridge, Cambridgeport, and North Cambridge) where most of the tenement housing was concentrated (Bunting 1971; Bunting and Nylander 1973; Downing, MacDougall and Pearson 1967; Krim 1977) (Map 12.1).

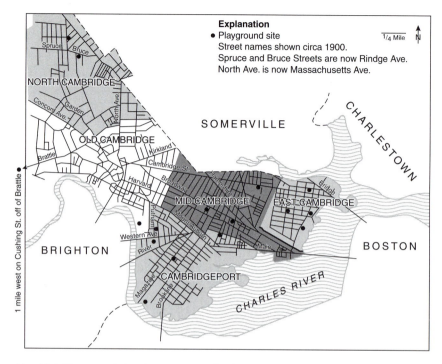

Map 12.1 Playground distribution in Cambridge, Massachusetts

From immigrant children to American boys and girls

Ideas of nationhood

The intended transformation of 'regular little street urchins' (Annual report, Cambridge Playgrounds 1909: 6) to respectable adults-to-be involved a further bifurcation along lines of gender. Although playground organisers usually discussed the problem of street children in undifferentiated terms, the model of the saved child was clearly constituted by two ideals. To this end, the Committee divided children by age and gender, establishing two types of playground: the first, to be supervised by a male instructor and attended by boys over 12 years old; the second, to be supervised by a female instructor and attended by girls of all ages and boys under 12.[3] In so doing, the activities taught to boys and girls were not only kept substantively different, but were identified with different spaces. To maintain binary gender roles, the coincidence of a feminine performance with a 'female' body must always be predictable and irrefutable. Given the necessary 'discontinuities that run rampant within heterosexual, bisexual, and gay and lesbian contexts in which gender does not necessarily follow from

218

sex' (Butler 1990: 135), maintaining physical distance between activities was vital to preclude the disruption of normative performances.

The political implications of this regime are significant. While gender is always already a political phenomenon, the performances upheld in playgrounds were directed toward specific socio-spatial consequences. Playgrounds for older boys were devoted solely to sports, most commonly baseball, although track and field athletics were also popular. Playground instructors considered baseball the ideal sport to correct wayward boys who were starved of a conventional American upbringing. Where else could a boy learn to obey authority, be loyal to a team to the point of personal sacrifice, to endure physical strain and enjoy communal pride? Playground organisers considered baseball the perfect activity to produce ideal future citizens from children whom, they feared, would otherwise form gangs on the streets and amuse themselves with activities considered far from socially desirable. 'The ideal team player was the blueprint for the ideal citizen' (Cavallo 1981: 89).

In contrast, girls were taught quiet, non-competitive activities like sewing, craft work, and knitting, or they learned song and dance routines. Rather than competitive team games in preparation for civil life, girls made miniature clothes for their dolls, crafted furniture for their doll's houses, and crocheted. Rather than being encouraged toward a future role in civic life, girls were prepared for domestic duty. Their role as care-givers, anticipating a future in domestic labour either paid or otherwise, did, however, contribute to a discourse of nationhood. Where the masculine nation stood for civic activity, the feminized nation stood for domesticity (Dowler 1998; Enloe 1989; McClintock 1993; Nash 1994; Radcliffe 1996).

The positioning of women outside of civil activities may be unsurprising at a time when women could not vote. As Sally Marston (1990) reminds us, the constitution of the new Republic did not aim toward the inclusion of 'all' Americans when it specified 'all citizens'. Women, slaves, native American and propertyless white men were not considered citizens and were therefore unproblematically excluded from formal membership of the citizenry (ibid). Despite this exclusion from the franchised nation, women's domestic role was drawn discursively into the symbolic nation. Drawing from Anne McClintock's (1993) claim that gendered national identity has a temporal dimension, where nostalgic images of the past are mapped onto femininity and progressive nation-building is represented by masculinity, Sarah Radcliffe (1996) argues that gendered national identity is also thoroughly spatial. Thus, girls' playground performances simultaneously imply a spatial directive to return to the home. Though their participation in various patriotic rituals (discussed later) clearly points to their symbolic inclusion in the nation, their contribution would be appreciated, literally, elsewhere.

Expression or performance?

Playground organisers did not separate male and female activities with the intention of preventing 'subversive bodily acts' (Butler 1990) *per se*. The illusion of a neat unity between corporeal acts and certain bodies was, however, effectively imposed by the segregation of play. Playground organisers endorsed what Butler (1990) calls the 'expressive model' of gender signification, which, in turn, legitimised the logical separation of girls and boys. Immigrant children, it was thought, had no access to normative examples of how best to behave. Although play instructors were committed to teaching 'correct' ways to children, they did not purport to be the sole influence on behaviour. Instead, instructors believed they were merely exhuming in-built inclinations from the child's soul. Never having been exposed to the normative regimes upheld in 'polite society', children could not know how to behave. With the right influence, however, their instincts to perform properly would emerge. Rather than acknowledge the regulatory framework that was responsible for inducing masculine and feminine performances, instructors advanced the notion that children were merely expressing their new-found and true identity.

The contradictions evident in the instructors' accounts of playground activities illustrate the tension between their faith in the expressiveness of identity and their concomitant desire to control the learning options available to children. The Committee expected instructors both to co-ordinate activities that would best train children and to reflect model characteristics themselves. The Committee write:

> too much cannot be said of the effect upon the boys of the personality of the director. Boys are quick to respond to the character that is over them . . . Whatever the director is, that is what the character of the playground will become.
>
> (Annual report, Cambridge Playgrounds 1909: 10)

Evidently, the influence exerted upon children was considered to be a significant component of children's transformation. Children must be able to witness the performances they too were expected to follow. First and foremost, then, children must be confronted with the correct alignment of 'sex' with gender performance. Thus, a female instructor, always unmarried, was responsible for younger children's playgrounds, and a male instructor was always placed in charge of older boys. Both were expected to provide immaculate renditions of their respective roles. Describing the style of instructor expected for boys' playgrounds, the following passage suggests a recognisably masculine demonstration:

In the employing of capable instructors, strong minded and physically able, for although and [*sic*] instructor may be very able . . . he may not have the back bone or the physical strength and endurance to force down mutiny and protect his own boys against the loafer and rowdy who will some times attempt to interfere.

(Report for Sargent out-door playground, Mr. Bruce 1909: 6)

Despite this manifest faith in 'external' forces, the Committee and instructors simultaneously regarded children as containing within them the impulses to act properly. Rather than coercing children into alien behaviour, it was expedient for instructors to assume that they were merely coaxing out a set of behaviours that were present but heretofore unliberated. The following passage is written by the instructor of a boys' playground. Initially, the instructor writes, the boys were 'offish':

Gradually, however, they began to feel a certain liking for me which grew at the close of the season to a tremendous devotion and personal loyalty. This feeling gives the instructor a golden opportunity to improve the living of some of its members with regard, for instance, to smoking, swearing, gambling, and cheap reading. As I said before, the moral virtues can be found in everyone of them and they can appreciate a noble act as well as anyone. Can they perform noble acts themselves? Only give them a director who will do his best to encourage such acts and who will be a continual example to them, and their inherent good-ness will crop out in all sorts of ways. It is not asking too much of a physical director to act as an example of good morals and good living for those boys in their formative period.

(Annual report, Cambridge Playgrounds 1909: 10)

Children, then, were written into their own normativity. Instructors argued that, though their influence was of critical importance to children's behaviour, they were not demanding children to do or be anything other than what came naturally to them. In so doing, the putative acts of gender were upheld and renewed in each child, while that process was simultaneously displaced onto the child's will. As Butler (1990) argues, it is the implied naturalness of this system that obstructs the politicisation of gender construction.

Playground performances

In this section I turn to the specific contours of gender performances. Although games and activities, and the particular methods of encouraging the proper ways

of enacting those, children were directed toward corporeal styles that constituted their gender.

The programme of play in boys' playgrounds was directed by the instructor's
desire to produce model citizens from immigrant boys. The team game provided
the stage on which their citizenly acts could be rehearsed for adult life. Baseball
was a means to encourage loyalty, both to other team members and to the
symbolic entity they represented, and provided a mechanism for encouraging
the protection of property. Boys' ability to successfully render these performances became proof of their viability as potential citizens; although generated
through a tightly scripted system, their maleness was taken to be the cause rather
than the effect.

The idea of belonging to a collective organisation was used by instructors to
encourage loyalty that could be transferred to the nation. The first separate playground for older boys was organised in 1904, two years after the first playground
was established on Pine Street, in East Cambridge. Over the next few years a total
of six playgrounds were established specifically for older boys. In addition to the
instructor who organised activities in each yard, the Committee employed a supervisor to oversee all playgrounds. Each playground had a baseball team who took
the name of their yard. In 1908, these teams were organised into a league, ensuring regular matches that culminated in a battle for the final cup. Not only did this
facilitate 'healthy' competition between boys, it also reinforced loyalty and alle-
giance, as each boy was now operating within the regulations of an organised community which he represented and fought for. In his general report, as supervisor
of all playgrounds, Mr. Candee writes of the successful realisation of 'team spirit':

> the boys who were loyal, in the end, were proud of it because their
> playground had proved to be something which it was a privilege to fight
> for. They came to realise that the interests of their one playground
> were not the only important questions, they began to feel that they
> were part of an organisation which had certain policies directing it and
> to be a worthy part of that organisation, those policies must be followed.
>
> (General Report, Mr. Candee 1909: 2)

Boys were also expected and encouraged to reassess their attitude towards
property. Instructors assumed that boys entered playgrounds with lax attitudes
towards property and embarked on a mission to instil boys with the necessary
impulse to honour others' possessions and protect their own and their community's belongings from outside attack:

> At the beginning of the year it was very hard to recover the apparatus,
> material, etc. at the end of the day's play. After the first two weeks

the boys were made monitors of their own goods, (which belonged to the league), and it was instilled into them that such materials of the Cambridge Playground League were their own and that the loss of such material would mean loss to them personally. This plan worked with great success, the boys taking hold of their work with great zeal, the fellows who in any way tried to steal the property of the League being punished speedily by their own play-mates.

(Annual report, Cambridge Playgrounds 1909: 4)

By this system, boys both performed their duty to protect and were also taught to recognise and punish behaviour that was classified as deviant. Thus, the extension of playground lessons in time and space was secured. Boys' performance as guardians of property, territory and honour was animated through their opportunity to foster such roles in team games. Furthermore, in the process of casting boys as the protectors of territory and property, girls were deftly allotted the position of the protected.

Cast in opposition to the heroic performance of boys as physically able defenders and loyal subjects, girls were to perform a patriotic narrative that contributed to, but did not collide with, masculine national identity. Narratives of the nation could sustain two gendered versions without disruption to feminine and masculine norms precisely because they were contained by different bodies. Rather than being trained for active participation in the civil realm, girls performed acts that were to embody the nation as a symbolic but separate entity, their bodies standing for, but remaining outside of, public sphere duties. Their participation in the nation would be from the home and would be represented by activities that serviced, supported, and cared for the nation's men and children (or if in domestic labour, the nation's affluent). Enacting their endorsed roles in playful ways, they were drilled in domestic crafts, sewing and knitting, child care, and for physical activity, were taught patriotic dances and anthems.

Daily activities in playgrounds for younger children were strictly time-tabled. A typical day might run as follows: activities would begin at 9:00 a.m. with folk dances and ring games, at 10:15 a.m. they would begin craft work, the afternoon session would begin at 2:00 p.m. with general sewing and from 3:30 to 4:30 p.m. there would be general play (Report of Cambridge Field, Miss Kitchin 1910). The games and dances that often began and ended the day were central to playground routines and were often the only performances that the public were encouraged to view directly.[4] On exhibition days, usually held at the end of the summer, girls performed aspects of their well rehearsed play for the surrounding community. Though some routines were less obviously marked by patriotic sentiments, the presentation was framed by jingoistic ceremony. With reference to an exhibition morning at Cambridge Field, the instructor writes,

223

'a large space in front of the balcony was roped off for the performance which started with an American Flag Drill, done to the tunes of patriotic airs' (Report of Cambridge Field, Miss Kitchin 1910: 14). Throughout the instructor's reports, they note 'patriotic songs' being sung not only for exhibitions but as part of the daily regime, including the *Star Spangled Banner*, *My Country 'Tis of Thee* and *America* (Report of Cambridge Field, Miss Kitchin and Miss Rea 1910). Girls could clearly embody the nation in their performance, which, when shown to the public, served to demonstrate not only their adoption of American pride but their peculiarly feminine rendition of it. Compared to boys' metonymic relationship with the nation, Joanne Sharp (1996: 99) writes, 'Women are not equal to the nation but symbolic of it'. Although, she continues, 'many nations are figuratively female . . . In the national imaginary, women are mothers of the nation or vulnerable citizens to be protected'. The feminine version of nationhood scripts girls as the harbingers of national pride, but their signifying performance defers democratic involvement to an active citizen from whom they are severed by bodily difference.

Despite the symbolic orientation of girls' performance of the nation, their role was not necessarily conceived of as being inactive. Their consignment to the home brought with it necessary reproductive and supportive roles. Girls were trained for a future of care-giving, the most important feature of which was reproductive. Girls' bodies, though able to represent the nation in ceremony, were more importantly to 'reproduce the bodies of the "body politic"' (Nash 1994: 237). Their maternal performance was to be given the space to develop in the playground. At the age of 12, boys left the younger children's playground to take their place on the baseball diamond. Girls, however, did not have an alternative provision. This is explained, in part, by the fact that they did not require a separate space to rehearse their future role. On the contrary, their performance depended upon the presence of younger children to enable their care-giving skills to be fine-tuned. At the age of 12, girls assumed the role of helper in younger children's playgrounds, becoming, according to the Committee, 'little mothers' (Annual report of Playground Pine Street 1904: 2).

The separation of boys' and girls' activities, as stated earlier, sustains the effect of inevitable gender performances. In the remainder of this chapter I turn to specific instances when the Committee and instructors directly compare boys and girls, pointing to a more explicitly articulated dissonance between gender expectations.

Mary Douglas (1966: 36) suggests that 'dirt is the by-product of a systematic ordering and classification of matter' and thus dirt is merely 'matter out of place'. Commonly understood as a polluting presence, dirt is undesirable; it represents confusion and jars against the 'cherished classifications' (*ibid.*: 37) of social order. Throughout Western history there has been a consistent effort to map dirt and defilement onto peoples and places that represent nonconformity (Sibley 1995). Dirt and amorality combine in similar ways. In both nineteenth-century Britain and America, immigrant and working-class populations have been targeted as populations in need of 'moral cleansing' through physical cleaning (Sibley 1995: 57). In playgrounds, cleanliness was employed to similar ends. Instructors used increased levels of cleanliness as a clear index of moral improvement in children. While this clearly incorporated a racialised discourse that plotted dirt against difference, the route to conformity again bifurcated along lines of gender. For boys, cleanliness caused few occasions for comment, and when it was recognised as a 'problem', it was solved quickly by 'clean hands and faces'. For girls, 'cleanliness' required their attention not just to scrubbed extremities, but to an aesthetic of appearance.

Generally, instructors used circuitous methods to induce (their notions of) cleanliness in children. They would often use clean hands and faces as a qualification for certain activities. For instance, when a travelling library was introduced, only children who could demonstrate suitable levels of cleanliness were permitted to borrow books (Annual report, Cambridge Playgrounds 1908). The style of cleanliness, however, assumed different shapes for boys and girls. The following statement reveals the disparity between the relative expectations for their improvement. Whereas boys' raucous behaviour was to be tamed by play, for girls, taming appearance was awarded higher priority:

> the boys never give me any trouble now and the girls are much more careful with their appearance. One has changed from a very dirty baby to a clean-faced child with hair nicely combed, coming every day ready for beads, with her needle to be threaded. This week she has come with shoes and stockings. This I am sure has come about by me noticing her and making much of her.
>
> (Annual report, Cambridge Playgrounds 1909: 7)

Rather than learning the correct way to behave, girls were, in addition, required to adopt particular ways of carrying and presenting their bodies (Plate 12.1[5]).

In addition to the cleanliness expected of children, instructors also endeavoured to keep playgrounds tidy. To that end, chores were allocated to children, both boys and girls, in an effort to cultivate 'improved' standards. However, for boys to perform an act that is traditionally designated feminine, it required reinterpretation according to a male script. Thus, to warrant boys' participation in domestic duties, instructors established 'cleaning brigades'. The Committee write: 'The boys are organised into a police force . . . She [the teacher] embroidered "S.P." for (sandgarden police) on their caps, and each season they made a tour of the yard, picking up papers and broken glass' (Annual report

Plate 12.1 Children lining up to receive apples. © Schlesinger Library, Radcliffe College

1907: 3). Whereas girls were expected to clean without trophies or uniforms, boys were offered an authoritative guise to safeguard their appropriation of a nominally feminine act. To have a 'male' body perform an act that usually serves to identify 'female' bodies, onlookers must be assured that his gender remains intact.

Conclusion

In this chapter I have argued that the regulation of gender through the 'surface politics of the body' (Butler 1990: 136) emerges through playground practices. In the playgrounds discussed here, the regulation of gender is facilitated by the isolation of taken-for-granted body types, and the simultaneous grafting of opposing performances onto each assumed type. The performances expected from girls and boys, through their play, were comprehensible within a larger social framework. Not only was sexuality made available to children in a heterosexist framework, but the roles expected from sexual 'opposites' defined political inclusion, reproductive responsibilities, and established norms of bodily presentation.

The performative, and therefore necessarily flexible, nature of gender is concealed by institutional commitment to 'expressive' notions of identity. The Playground Committee evidently subscribed to and promoted essential notions of gender; however, their conviction was destabilised by their concurrent effort to impose particular behaviour on purposefully differentiated bodies. If gender is to be performed differently, 'outside the restricting frames of masculinist domination and compulsory heterosexuality' (Butler 1990: 141), then the performative nature of gender needs first to be recognised. In this chapter I have attempted to outline some of the practices used in learning environments that limit the ways in which gender can be enacted without drawing attention to itself. By adopting learning practices that impose normative limits on possible performances, conservative gender politics are reproduced. More preferable, perhaps, are environments through which alternative renditions of gender are not only possible but come without punishment.

Acknowledgements

Many thanks to John Paul Jones, Mitch Rose and Rich Schein, for taking the time to read and comment on earlier drafts of this paper. Thanks also to Sarah Holloway and Gill Valentine for their guidance and editorial comments. In addition, I would like to thank Carl Dahlman and Rebecca Sohmer who kindly contributed their cartographic skills. I am grateful to the Schlesinger Library, Radcliffe College, for permission to publish material from their manuscript and pictorial collections.

Notes

1   This figure includes all play facilities, both supervised and unsupervised. A more modest estimate for cities with supervised playgrounds puts the number at 67 in 1908, rising to 102 in 1909 (Rainwater 1922).
2   This report, given by Mrs. Almy of the MCC and later the Playgrounds Committee, followed a visit to New York City where she inspected vacation schools established by the New York Association for the Improvement of the Poor. The apparent success of these schools prompted her suggestion that Cambridge would benefit from a similar provision.
3   Although younger boys shared space with girls of all ages, I refer to these playgrounds as 'girls' playgrounds' as my focus is on their activity. Feminised activities dominated these playgrounds, but activities for boys and girls remained distinct.
4   Public performances of playground routines did take place, but girls' exposure to the public was strategically contained for fear of 'inappropriate' viewing (see Gagen in press).
5   There are no specific records to explain the content of photographs. However, the reports describe various occasions where children lined up to receive flowers or fruit from instructors.

References

Aitken, S.C. (1994) *Putting Children in their Place*, Washington, DC: Association of American Geographers.

Boyer, P. (1978) *Urban Masses and Moral Order in America, 1820–1920*, Cambridge, Mass.: Harvard University Press.

Bunting, B. (1971) *Survey of Architectural History in Cambridge, Report Three: Cambridgeport*, Cambridge, Mass.: Cambridge Historical Commission.

Bunting, B. and Nylander, R.H. (1973) *Survey of Architectural History in Cambridge, Report Four: Old Cambridge*, Cambridge, Mass.: Cambridge Historical Commission.

Butler, J. (1988) 'Performative acts and gender constitution: an essay in phenomenology and feminist theory', *Theatre Journal* 40, 4: 519–31.

—— (1990) *Gender Trouble: Feminism and the Subversion of Identity*, New York: Routledge.

—— (1993) *Bodies That Matter: On the Discursive Limits of "Sex"*, New York: Routledge.

Cavallo, D. (1981) *Muscles and Morals: Organised Playgrounds and Urban Reform, 1880–1920*, Philadelphia: University of Pennsylvania Press.

Douglas, M. (1966) *Purity and Danger: An Analysis of the Concepts of Pollution and Taboo*, London: Routledge.

Dowler, L. (1998) '"And they think I'm just a nice old lady": women and war in Belfast, Northern Ireland' *Gender, Place and Culture* 5, 2: 159–76.

Downing, A.F., MacDougall, E. and Pearson, E. (1967) *Survey of Architectural History in Cambridge, Report Two: Mid Cambridge*, Cambridge, Mass.: Cambridge Historical Commission.

Enloe, C. (1989) *Bananas, Beaches and Bases: Making Feminist Sense of International Relations*, Berkeley: University of California Press.

Environment and Planning A (in press) 'From crib to campus: institutional spaces and gender/sexual identities', special issue ed. S.C. Aitken.

Gagen, E.A. (in press) 'An example to us all: child development and identity construction in early twentieth-century playgrounds', special issue *Environment and Planning A*.

Goodman, C. (1979) *Choosing Sides: Playground and Street Life on the Lower East Side*, New York: Schocken Books.

James, A., Jenks, C. and Prout, A. (1998) *Theorising Childhood*, New York: Teachers College Press.

Krim, A.J. (1977) *Survey of Architectural History in Cambridge, Report Five: Northwest Cambridge*, Cambridge, Mass.: Cambridge Historical Commission.

Marston, S.A. (1990) 'Who are "the people"?: gender, citizenship and the making of the American nation', *Environment and Planning D: Society and Space* 8: 449–58.

McClintock, A. (1993) 'Family feuds: gender, nationalism and the family', *Feminist Review* 44: 61–80.

Nash, C. (1994) 'Remapping the body/land: new cartographies of identity, gender, and landscape in Ireland', in A. Blunt and G. Rose (eds) *Writing Women and Space: Colonial and Postcolonial Geographies*, New York: Guilford Press.

Platt, A.M. (1977) *The Child Savers: The Invention of Delinquency*, Chicago and London: University of Chicago Press.

Ploszajska, T. (1994) 'Moral landscapes and manipulated spaces: gender, class and space in Victorian reformatory schools', *Journal of Historical Geography* 20, 4: 413–29.

Radcliffe, S.A. (1996) 'Gendered nations: nostalgia, development and territory in Ecuador', *Gender, Place and Culture* 3, 1: 5–21.

Rainwater, C. (1922) *The Play Movement in the United States*, Chicago: University of Chicago Press.

Rivlin, L.G. and Wolfe, M. (1985) *Institutional Settings in Children's Lives*, New York: John Wiley and Sons.

Ruddick, S.M. (1996) *Young and Homeless in Hollywood: Mapping Social Identities*, New York: Routledge.

Sharp, J.P. (1996) 'Gendering nationhood: a feminist engagement with national identity', in N. Duncan (ed.) *BodySpace: Destabilising Geographies of Gender and Sexuality*, London and New York: Routledge.

Sibley, D. (1995) *Geographies of Exclusion: Society and Difference in the West*, New York: Routledge.

Valentine, G. (1996) 'Children should be seen and not heard: the production and transgression of adults' public space', *Urban Geography* 17, 3: 205–20.

—— (1997) '"Oh yes I can". "Oh no you can't.": children and parents' understanding of kids' competence to negotiate public space safely', *Antipode* 29, 1: 65–89.

Ward, D. (1989) *Poverty, Ethnicity and the American City: Changing Conceptions of the Slum and the Ghetto, 1840–1925*, Cambridge: Cambridge University Press.

Manuscript collections cited

Annual reports for Pine Street Playground, 1904; Cambridge Playgrounds, 1907, 1908 and 1909, in Mother's Club of Cambridge papers, box 1, folder 1, Schlesinger Library, Radcliffe College, Cambridge, Massachusetts.

General Report, Mr Candee, 1909, in Helen Jackson Cabot Almy papers, folder 5, Schelsinger Library, Radcliffe College, Cambridge, Massachusetts.

Mrs. Almy's report to the Mother's Club of Cambridge on Vacation Schools, n.d., in Helen Jackson Cabot Almy papers, folder 1, Schlesinger Library, Radcliffe College, Cambridge, Massachusetts.

'Playground Facts' leaflet number 2, published by the Playground Association of America, 1909, in Helen Jackson Cabot Almy papers, folder 8, Schlesinger Library, Radcliffe College, Cambridge, Massachusetts.

The Record, 9 February, 1909, No. 23, in Helen Jackson Cabot Almy papers, folder 7, Schlesinger Library, Radcliffe College, Cambridge, Massachusetts.

Report of Cambridge Field, Miss Rea, 1910; Report of Cambridge Field, Miss Kitchin, 1910, in Helen Brooks Collection, box 6, folder 218, Schlesinger Library, Radcliffe College, Cambridge, Massachusetts.

Report for Sargent out-door playground, Mr. Bruce, 1909; General Report, Mr. Candee, 1909, in Helen Jackson Cabot Almy papers, folder 5, Schlesinger Library, Radcliffe College, Cambridge, Massachusetts.

13

WALK ON THE LEFT!

Children's geographies and the primary school

Shaun Fielding

Introduction

At the beginning of the 1990s, the Social and Cultural Geography Study Group (SCGSG) of the Institute of British Geographers (as it then was) began to consider the ways in which 'the social' and 'the cultural' were being re-conceptualised and re-positioned within and through human geography (see the articles in Philo 1991). A key outcome was the apparent reconnection of human geographic discourse to post-Kantian moral philosophy with its common moral talk:

> The 'common sense' moral assumptions that all of us routinely make in our everyday lives in order to establish what should be done, who should be trusted, where we should go and so on.
>
> (SCGSG 1991: 14).

A serious consideration of common moral talk, in itself and in relation to wider moral systems (such as schooling in the case of this chapter) could, it was suggested, interrogate the concept of *moral geographies* with its three interrelated strands. One strand focused upon the need for geographers to reflect upon their own morality, their own sense of right and wrong in the values, practices and beliefs that they brought to their work. A second strand considered the geography *of* moralities that are reflected in the moral assumptions that people in different parts of the world make about binary oppositions such as good and bad, just and unjust. A final strand, and the focus of this chapter, examined the geography *in* everyday moralities, the belief that our moral assumptions and arguments have an in-built consideration for spaces and places (SCGSG 1991).

This idea of an in-built geography to moral views within an educational context was driven home to me during my first days as a geography teacher on

Merseyside. I had organised my classroom so that the tables made a large rectangle in the middle of the room, with the material resources laid out around the classroom's perimeter. This was designed to facilitate better co-operation and collaboration among the children, and to foster this, I was going to allow them to sit in friendship groupings. On my first day, the head of department came in and told me to rearrange the classroom into rows and to seat the children in alphabetical order. A few days later, myself and a Year Seven class (11–12-year-olds) walked round the school to familiarise ourselves with its geography. During this, we met the headteacher who 'told us off' because we were walking on both sides of the corridor and not the down the left-hand side, and because we were walking in groups and not in pairs.[1] Unprofessional practices aside, what struck me was that the school, its beliefs and practices was a 'hot bed' of moral geographies – of moral codes about how and where children ought to learn and behave – and that the ways in which these were played out by the children (the children's geographies) were of significant importance.

Consequently, this chapter will extend this theme by identifying some of the moral geographies within a selection of English primary schools and the children's geographies that are constructed out of these. A useful connection here is provided by the work of the German sociologist Norbert Elias (1978, 1982, 1987) and those who have demonstrated the importance of his work on the historical construction of the self (Ogborn 1991, 1995).

Civilising and schooling

In his works *The Civilising Process Volumes I and II* and *Involvement and Detachment*, Elias foreshadows developments in social theory and geography by highlighting the importance of the training and disciplining of the body and the self and then connecting this to state formation. For Ogborn (1991, 1995) Elias is important because his theory of the self suggests ways in which its conceptualisation can become spatialised into 'inside' and 'outside' at certain periods and places. Furthermore, the central concept of figuration as a way of thinking about societies, institutions and individuals is, Elias suggests, an extended spatial metaphor. Elias signposts 'a long term change in human affect and control structures and in the regulation of desires and emotions' (Ogborn 1991: 79). People, he suggests, have experienced a tendency towards an increased tightening and differentiation of controls and a growing threshold of aversion, if you like, an increasing morality to their own geographies. He illustrates this through an historical analysis of social manners and etiquette books whose purpose was to inscribe 'apt' behaviour. In this, he detects a shift to a moral strictness with a closer control of bodies and an increased observation and sensitivity towards oneself and others. In addition:

A vital part of this transformation is a spatial separation, a shifting behind the scenes of things previously accepted . . . and a suppression of cruelty and violence in everyday life except in certain specific conditions strongly controlled by the state.

(Elias 1982, in Ogborn 1991: 80)

One of the techniques identified by Elias was the way that external restraints and prohibitions are internalised by people so that their behaviour is managed by varying feelings of disgust and aversion. By transferring this technique to a particular setting such as a primary school, these social prohibitions could be seen as a way of regulating and situating oneself within its social hierarchy. Those who are unable to internalise such restraints are more likely to be excluded or marginalised, a fact which is regularly communicated to them by the speech and actions of 'other' people. Whilst a body of work which expresses these or similar sentiments has yet to surface in education research, some historical geographers and architects have incorporated ideas about bodily propriety, manners, discipline and morality in the context of education/schooling in their research and these warrant further attention.

Children and historical geographies of schooling

Until recently, the main insights into children's geographies in schools have been provided by historical geographers. Gagen (1998; see also Chapter 12), for example, describes the role of playgrounds in turn-of-the-century Massachusetts as a way of 'getting hold of the children'. As she argues in Chapter 12, although not schools *per se*, these playgrounds provided environments for 'sites of transformation'. Children represented the potential solution to the lost tradition and morality brought by the onset of modernity. According to the reformers, the training of children through physical order would engender social transformation and lead to the flowering of hitherto invisible (and morally acceptable) inner identities (Gagen in press). The playgrounds and their practices acted as an ideal social control/supervisory mechanism to bring children out from the urban tenements so as to demonstrate visually the restorative properties of 'moral reform'. They also functioned both directly and indirectly to inculcate and reinforce gender/bodied identities and to homogenise their charges, to 'embody America' and so eradicate their foreign traits. Gagen concludes her chapter by pointing out that the playgrounds were a programme of social control with explicit moral geographies, camouflaged by a romanticised ideology of learning by play.

Similar ideological struggles for the hearts and minds of working-class children took place in the reformatory schools of Victorian Britain as studied by Ploszajska (1994). She focuses upon two schools (one for boys: Farm School,

Redhill; and, one for girls: Red Lodge, Bristol). Their central premise, she argues, was to create new institutional programmes that recognised the need for regularity, discipline and surveillance but which minimised the need for external physical coercion. This was thought to bring about the internalisation of moral standards amongst those members of society perceived to be, somehow, deviant or socially threatening (Ploszajska 1994: 413).

Farm School was based in part on the path-breaking reformatory at Mettray, France, where boys were divided into 'families' of about forty children. Each 'family' lived in separate houses supervised by a master and two senior boys, which were situated around a central square so as to facilitate constant surveillance and to minimise potential disruption. In contrast to Mettray, at Farm School, the reformatory houses were located randomly on the site so as to create distinct smaller schools and instil competition between each 'family'. The discrete communities only came together when in Chapel. In order to ensure effective surveillance, the family houses were organised around carefully planned large-scale and open-plan day rooms and dormitories which enabled 'the family' to operate together and which deliberately cut out concealed corners and potential hiding places (Ploszajska 1994: 418). At Red Lodge, incoming girls had to sleep in a separate space before they could join the main school. This was seen as a means of purification, or of distilling undesirable thoughts and actions so that they could not influence the girls already into their 'reforming'. The dominant discourses here were of 'domestic order', of 'family feeling' and of the school as a 'happy home'.

As these (all too brief) historical examples demonstrate, there are powerful moral and spatial forces at play in the design, organisation and management of schooling, which act as modes of regulation for 'appropriate' social behaviour. In the rest of this chapter I will concentrate upon the ways in which similar forces are sculpted into contemporary primary schools in the UK. The empirical material was collected from ethnographic research in several primary schools in the West Midlands.[2] The ethnographies were bolstered by one-to-one semi-structured interviews with head, deputy-head and class teachers and by informal discussions and conversations with teachers and children (mainly in Years Five and Six – 10- and 11-year-olds).[3] The names of the schools and the identity of the teachers and children have been changed to preserve anonymity.

Children and contemporary geographies of schooling

The recent burgeoning of literature around children and young people's geographies has been at pains to point out children's competency as social actors. The importance of children as negotiating subjects and active participants in and over space, despite unequal institutional power relations and control, is at the forefront of many research activities, not least in this and previous collections (Skelton and Valentine 1998). In UK primary schools, the extent to which this agency is realised is largely dependent upon the structuring of the teaching, learning and management within the school, which is in turn constructed through the moral beliefs and practices of the governors, headteacher, teachers, learning support assistants, the Local Education Authority (LEA), the Office for Standards in Education (Ofsted) and central government. In schools, which articulate and display a relatively transparent approach to the organisation and delivery of teaching and learning, clear messages are sent to the teachers about what it means to be a good teacher and to the children about what it means to be a good learner in their school/classroom and what social behaviour is expected from them both. These messages are transmitted through discourses and inter-action (between headteacher and teacher, between teachers and teachers, between teachers and pupils and between pupils and pupils), through the wall displays, through bodily movement and posture and through the school rules. In some schools, these rules are no longer a list of do's and don'ts, but a series of nego-tiated positive expectations displayed as a formal pupil charter or mission statement. For example:

> We will always aim to improve upon our previous best.
> We understand the importance of helping people to achieve all that they can.
> We will treat people with the respect that they deserve.
> It is important to understand and respect the differences between people.
> (Extracts from the Mission Statement of Howley School)

To follow Elias's ideas of bodily propriety into the classroom, I observed that these moral codes and messages about being good learners were embodied by many children in their day-to-day learning activities. Their body postures infer a willingness to collaborate (they avoid putting their arms across themselves, they sit back and look relaxed), there is very little pointing, leaning over or shouting, there is a greater fluidity of movement around the classroom (of both people and equipment), and a greater amount of 'on task talk' within groups. Likewise, there is an 'effective' moral policing of classroom activity by the chil-dren, such as when unacceptable forms of learning or behaviour present themselves. Here, they are quick to find closure with that person (in most cases through the aversion discussed by Elias) or they negotiate ways through this, by using their bodies to display dismay, by introducing pointing and leaning as they talk with the person and the other people on the table, or they then involve other people (such as the teacher) if needs be.

INTERVIEWER: What happens in your group if you cannot work together?

ANDREA: We try and work it out first because Miss has told us that that is the best way.

DAVINA: Sometimes I just go and sit in the reading area because I get upset . . .

ANDREA: Yeah, especially when Kevin is messing about?

INTERVIEWER: What do you do when Kevin messes about?

TARA: We ask him to stop, and sometimes he stops, if not we'll go and ask Bobby to come over because Kevin respects Bobby . . . if not, we'll tell Miss.

ANDREA: But we always give him a chance because we tell him we are going to tell Miss.

TARA: Yeah, you know he has like three strikes and out.

(informal conversation with Howley primary school children)

In schools where there appears to be ambiguity in the forms that the delivery of teaching and learning take, these messages become somewhat distorted, especially by the children in the classroom. Bodies become more ambiguous as hands and arms are used as barriers around their work, but also to bring people into their work; bodies are turned away from people, but later open up to offer or seek co-operation; volume rises and falls more sharply; classroom movement becomes more disjointed or haphazard and so on. In one school, the class teachers began to present their own moral codes to the children as an antidote to the 'ambiguity of expectation' further up the school hierarchy. This effectively created many small pockets of different moral geographies around the school with no overriding steerer. Consequently, when these came together in collective activities (such as at break and lunch time, or during assembly and hymn practice), they clashed with each other, both in terms of what the children and teachers understood to be acceptable learning and behaviour, but also how they would 'police' and 'sanction' what they believed to be unacceptable behaviour.

> As a new teacher in this school I get no sense of any educational leadership, everyone seems to be doing their own thing to their own set of rules. Don't get me wrong, professional autonomy is vital, but that needs to be in the context of an overall vision which we simply do not have . . . This makes it very confusing for me, so imagine what it is like for the children.
>
> (interview, Year Five teacher, Downdale school)

I agree that advocating the imposition of collective, centralised moral codes and geographies in primary schools stultifies professional autonomy (a fact which the current administration seems oblivious to). Nevertheless, an overt moral 'framework' within which children and teachers are free to establish the best forms of teaching and learning in their particular space *appears* to generate not just 'successful' teaching and learning (though this is such a loaded term), but throws up a whole suite of differing classroom and children's geographies that teachers and children can reflect back on to inform future activity.

This was the case in Park Lane school, a large (three-form entry) primary school in a lower-middle-class catchment area, with an above-average percentage of children from ethnic minorities. The educational ethos of senior management encouraged professional autonomy within the rubric of a series of common moral aims. Here, the flexibility of human and material resources was a key organisational principle. Most of the junior classes (7–11-year-olds) were organised into sets for Maths and English and the three class teachers in each year group would take one set each. In addition, key senior teachers were given the role of 'floaters', non-specific class-based teachers who would teach in each class in the school to allow the class teachers time for planning and administration. The timetable was organised so that these 'floaters' taught the same subject throughout the school (science and art in this case). This organisation required that teachers share their classroom space with other teachers and that the children use the classroom space according to the differing teaching styles that were being offered to them by these different teachers.[4] Whilst in theory, this is an innovative use of resources for primary schools, tensions did arise between the different moral geographies that different teachers bring to the same physical classroom space and which the children had to adapt to and negotiate with in their own classroom geographies.

This was particularly the case in one Year Five class (9–10-year-olds). The class teacher Wendy was a young, small, white woman in her third year at the school. She dressed casually for work and, it was rumoured, was being fashioned for promotion.[5] Her pedagogic practices elicited very different classroom geographies from her children when compared with the pedagogy of David, the floating teacher for science. David was very tall, in his late forties, with a deep booming voice, and he had been at the school upwards of fifteen years and always wore a suit and tie.

According to the headteacher, Wendy's pedagogic practice reflected the 'best practices' of teaching and learning. I would describe it as a *seductive pedagogy*, where the children are seduced into learning scenarios through positive reinforcement, even pacing, an inclusive tone of voice and the use of modal verbs rather than commands.[6] Moreover, this pedagogy appeared to open up the classroom space for the children's learning. It encouraged their movement around and also out of the classroom in the search for knowledge and understanding. It enabled a fluid use of personal and classroom space, which paved the way for the interconnection of her own and the children's classroom geographies.

Furthermore, it appeared that this practice had created a kind of moral economy for the classroom, where children were allowed to broker for or exchange advice from her and other children on the understanding that they would reciprocate this either in this lesson or later on during the day, or they would help them with other school-related activities – a kind of LETS (a local economic trading system) for the classroom. Wendy describes these classroom dynamics as *orderly disorder*.

> I like that there is order within the disorder, the children know what they have to do and how to achieve the understanding and they know not to put themselves or the other children off task, so although there is all this 'messy' activity, there is a underlying purpose to it . . . I have to stress that we have worked very hard at getting this right and it only works because we have developed such a good relationship . . . There is no way that we could do this in September, I would have been locked up by Christmas.
>
> (Interview with Year Five teacher, Park Lane School)

This contrasted with David's pedagogic practice which, although different to Wendy's, still fell within the moral framework set out by the school (though not necessarily within the parameters championed by the head). I would describe it as a *reductive pedagogy*, one which was much more formal, hyper masculinist, loud and heavily centred around classroom control and classroom order, one which David himself embodies with his own body language, posture and clothes. The impact of this form of delivery upon the classroom and the children was that it appeared to close off the spaces for learning, restricting the children's use of their moral economy as they were required to undertake more formal individual learning styles. On reflection, this created a much smaller classroom, even though it was the same physical space. This reduction in the classroom space was also evidenced in the children's geographies, not least in their bodies. They sat scrunched up, leaning over their books, heads in their hands, or in front of their faces. Their movement and their communication was restricted, so that peer evaluation was negligible. It was as if there was some physical barrier hemming them in on all sides, like they were in some moral enclave.

It is perhaps easier to demonstrate these dynamics through some classroom maps. These examine the different 'geographies' displayed by one child (Rebecca) during a history lesson with Wendy and a science lesson with David. Rebecca was a white, working-class girl, who was considered to be helpful and a popular child within the class. She was of average attainment/ability, but was very good at sport and gained a lot of respect and kudos for this, especially from the boys. Map 13.1 illustrates the layout and the seating arrangements of the classroom

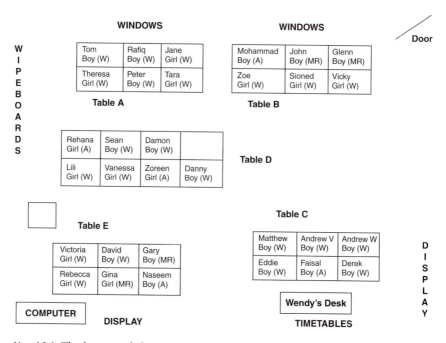

Map 13.1 The layout and the seating arrangements in Rebecca's classroom

and this does not change during either of the lessons. The children are seated according to their reading ability (which is an increasingly common practice in primary schools, particularly with the onset of the literacy hour). Map 13.2 illustrates Rebecca's 'learning activity' – her patterns of interaction and movement – during Wendy's history lesson. Map 13.3 represents that same 'learning activity' during David's science lesson. The dashed arrows represent movement to another child (double arrows indicate that this is reciprocated at some point). The bold lines indicate communication with another person (again, a double arrow indicates that this is reciprocated). The bigger the arrow, the greater the amount of communication or movement. On the Maps themselves, MR refers to mixed race children, A refers to Asian children and W refers to white children.

Although this a crude analysis, one can see the dramatic differences between Rebecca's 'learning activity' under these differing types of pedagogy. In Wendy's lesson, Rebecca is free to move to other tables and other children are free to come over and ask her for advice and guidance. She co-operates with boys and girls of differing ability and has a special relationship with Sioned, who has a very low reading age but is the best netball player in the year group: Rebecca helps Sioned with her work and Sioned helps Rebecca with her netball shooting.

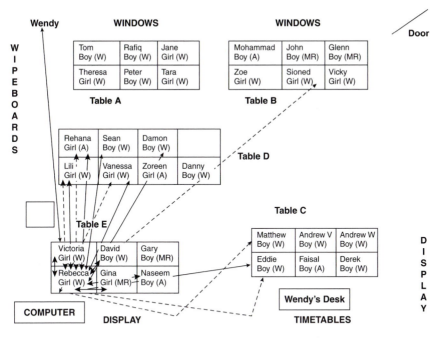

Map 13.2 Rebecca's 'learning activity' during Wendy's history lesson

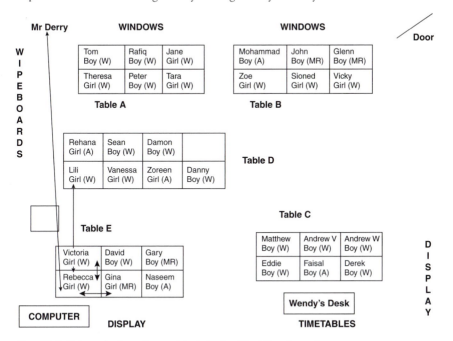

Map 13.3 Rebecca's 'learning activity' during David's science lesson

With the exception of Sioned, Rebecca's learning activities are focused onto what one might call a 'proximal space', that which is close to them. Wendy explained that after she and the class had got to understand who talked and worked with whom, where and when, they were able to develop a seating pattern that maximised the opportunities for everyone's learning, but which also minimised the potential for disruption – a fact upon which a school inspection team would look favourably.[7] In contrast, Rebecca's learning activity under David's pedagogy is much more restricted. Her movement between the tables and over her own table has been eradicated. She communicates only with the girls on her table as a kind of shelter from the hyper-masculinity of David's style (one has severely restricted the co-operation between boys and girls observed under Wendy).

While I would not defend David's pedagogy with its in-built inflexibility to different classroom situations and children's geographies, I have a partial expla-nation. As a 'floater' David visits at least four different classrooms a day, each with different spatial configurations that are designed around the specific peda-gogic style of that class teacher and their children's geographies. One might argue that David's reaction to these differing environments is to develop a generic, 'least cost scenario pedagogy' where classroom control is at a premium. Although this may hinder the learning activity of many of the children in these classes, I think David feels that at least every child will be on task and be able to learn something.[8] In addition, it is important to note that this is an approach that some school inspection teams look upon favourably given the appropriate contexts.

So, what can I offer by way of explanation regarding the children's reactions to these differing learning scenarios and their ensuing geographies? Whilst it would be fascinating to map out the 'learning activities' of other children and their different classroom geographies (such as Damon, a very clever boy, but a poor collaborator; or Derek, the naughty boy), space restricts my focus to just Rebecca. Rebecca's sense of being 'hemmed in' as described above is a partic-ularly difficult situation for a 'good co-operator' like her to either negotiate herself out of, or to destabilise. She is able to register her disapproval by inten-sifying the communication and co-operation across the table with herself, Gina and Victoria (and to a lesser extent, beyond the table with Lili). I read this as an attempt to carry out the same amount of learning activity, but over a much more compressed space. In a number of instances, this intensified collaboration became too much for that space, it simply had nowhere else to go, and it 'spilled over', manifesting itself as what David perceived to be disruptive behaviour and thus a threat to the moral geography/order of the classroom. 'We get into trouble with Mr Derry [David] more because we talk to much' (comment from Gina, Park Lane school, transcribed from field notes).

To a certain extent, the girls were able to manage David's intervention by claiming that they were talking about work, but they never asked to be able to go and ask other children (because they knew they would not be allowed to). In a couple of lessons, Rebecca managed to negotiate herself out of the classroom by going to the toilet and by getting some things from her bag. The tensions that Rebecca and her friends experienced were also reflected in similar behaviour by other girls in the class. Indeed, it seemed as if some of Rebecca's actions were to deflect David's attention away from another group 'in the same boat' and that those girls would do the same for them – a kind of counter moral geography, or a justifiable immoral geography. In contrast, the boys were extremely passive and never attempted to help the girls, largely (I presumed) because they appeared to be content with the pedagogy given to them by a male teacher in science (which they perceived to be a 'boy's subject').

In the context of those children, in that space, at that time, I would suggest that David is delivering a 'paradoxical pedagogy'. In order to manage and control the classroom space and the children's geographies in that space, he subjects those children to what he perceives to be a moral and ordered pedagogy. However, the order he creates actually leads to disorder (most significantly by the girls and those who are good collaborators and co-operators), which further disrupts and affects the children's learning.[9]

By way of a conclusion, it can be seen that the moral codes of this primary school classroom and their interpretation by different class teachers into a set of pedagogic practices, combined with the individual agency of children, contribute to the co-construction of a multiplicity of children's geographies that operate over one particular classroom space. The fact that I have only managed to scratch the surface of the children's geographies in that classroom, let alone in that school or others in the LEA, says a great deal about the level and the complexity of understanding that is required in order to bring such concepts and issues to the fore. One can appreciate how fraught the situation is for headteachers if they are to manage 'effectively' the teaching and learning in one primary school with all its different spatial settings, different pedagogical practices and different children's geographies. Nevertheless, even though children continue to negotiate their agency and to construct their geographies in increasingly sophisticated ways, central government has failed quite spectacularly to grasp the nettle of this sophistication. Instead, they have decided to subject children to a 'dumbing down' of their 'learning activity' through monolithic, prescriptive and centralised forms of learning for two hours a day, every day, under the guise of the literacy and numeracy hours. Not content with this, the Department for Education and Employment (DfEE) and Ofsted are encouraging education research to follow a similar 'dumbing down' process and to concentrate upon delivering 'evidence based research', which by definition neglects to engage sufficiently with theory. Naturally, this will be resisted on many fronts, but I feel that education researchers alone cannot carry the burden of this endeavour, and this is why geography and geographers are so important to the current cause of education and education research.

Geography and education: an interface or an interference?

As a geographer desperately trying to cling to that identity within education research, I am convinced that there is a huge potential for investigating and considering the role of space in schools and for a greater understanding of the dynamics of children's geographies so as to improve pedagogic practice. More importantly, geographers have an increasingly important role to play in helping education researchers and professionals to realise this. Such an agenda would pave the way for innovative and exciting research that interrogates the geography *in* education and that integrates our considerable understanding of the complexities of children's lives in and outside primary schools. Finally, given the recent fracas about education research, particularly the disparaging remarks about post-structuralist's tendency towards the 'adulation of great thinkers' (see Tooley 1998, and the responses in Bassey 1998a, 1998b; BERA 1998), there is a real danger that education will further distance itself from exciting *and* challenging research. I have faith that the growing 'intellectual alliance' between education researchers and geographers will enable people to consider the interconnections between space, identity, schooling and pedagogy. I am confident that such an alliance will bear fruit not just in academic circles, but in the corridors of policy-makers and decision-takers where it is vital that the complexities discussed in this chapter are woven into education policy.

Notes

1 Later in the year the head ordered all those teachers with Year Seven and Eight classes to spend the first lesson walking the children around the school so that they and the others watching could learn the correct way of conducting themselves in school.
2 These projects are: Teacher's Work: Changing Professionalism in a Restructured Public Service (ESRC No: R000233685) – I would like to acknowledge the support of the ESRC and of the project directors, Mairtin Mac an Ghaill (University of Sheffield) and Martin Lawn (Westhill College, Birmingham); and Enriching Learning: Developing the Primary School Interface Through Cross-Curricular Themes – I would like to acknowledge the support of Newtown and Ladywood Task Force on this project.
3 All of the quotations from discussions are written up from original field notes and thus total accuracy cannot be guaranteed.

4 It is important to note that other teachers such as Special Educational Needs or English as an Additional Language could also be in the classroom at the same time, as could learning support assistants and parent helpers.

5 A year later, Wendy was promoted to the head of English throughout the school.

6 Here, I wish to point out that some of the following concepts originally sprang from separate and ongoing research on gender and learning with Harry Daniels (Birmingham University) and Angela Creese, Valerie Hey, Diana Leonard (Institute of Education) and Marjorie Smith (Newham Psychological Services) and I would like to acknowledge their part in the development of these concepts in those particular contexts, which will be published at a later date.

7 For example, she kept Derek and Tom as far apart as she possibly could because they were 'at each other's throats all the time' (see Map 13.1).

8 As David declined to be interviewed, these ideas are speculative.

9 However, I would emphasise that I see this not as 'bad teaching' *per se*, because David's style generates different and productive geographies in other classes – it just does not suit the dynamics of that particular class.

References

Bassey, M. (1998a) 'Educational research: under investigation but robustly on the march!', *Research Intelligence: The Newsletter of the British Educational Research Association* (BERA) 64: 1.

Bassey, M, (1998b) 'Ofsted offside, but watch out for the IES review', *Research Intelligence: The Newsletter of the British Educational Research Association* (BERA) 65: 1.

BERA (British Educational Research Association) (1998) 'BERA's immediate response to the Tooley Report', *Research Intelligence: The Newsletter of the British Educational Research Association* (BERA) 65: 1.

Elias, N. (1978) *The Civilising Process. Volume One: The History of Manners*, Oxford: Blackwells.

Elias, N. (1982) *The Civilising Process. Volume Two: State Formation and Civilisation*, Oxford: Blackwells.

Elias, N. (1987) *Involvement and Detachment*, Oxford: Blackwells.

Gagen, E. (1998) 'Spaces of transformation: the (re)enforcement of gender ideals in turn-of-century playgrounds', paper presented at the Annual Conference of the Royal Geographical Society (with the Institute of British Geographers), Kingston University.

Gagen, E. (in press) 'An example to us all: children's bodies and identity construction in early twentieth century playgrounds', *Environment and Planning A*.

Ogborn, M. (1991) 'Can you figure it out? Norbert Elias's theory of the self', in C. Philo (comp.) *New Words, New Worlds: Reconceptualising Social and Cultural Geography*, Lampeter: Department of Geography, University of Wales at Lampeter.

Ogborn, M. (1995) 'Knowing the individual: Michel Foucault and Norbert Elias on Las Meninas and the modern subject', in S. Pile and N. Thrift (eds) *Mapping the Subject: Geographies of Cultural Transformation*, London: Routledge.

Philo, C. (comp.) (1991) *New Words, New Worlds: Reconceptualising Social and Cultural Geography*, Lampeter: Department of Geography, University of Wales at Lampeter.

Ploszajska, T. (1994) 'Moral landscapes and manipulated spaces: gender, class and space in Victorian reformatory schools', *Journal of Historical Geography* 20: 413–29.

Skelton, T. and Valentine, G. (eds) (1998) *Cool Places: Geographies of Youth Cultures*, London: Routledge.

SCGSG (Social and Cultural Geography Study Group) (1991) 'De-limiting human geography: new social and cultural perspective', in C. Philo (comp.) *New Words, New Worlds: Reconceptualising Social and Cultural Geography*, Lampeter: Department of Geography, University of Wales at Lampeter.

Tooley, J. with Darby, D. (1998) *Educational Research: A Critique*, London: Office for Standards in Education.

14

'OUT OF SCHOOL', IN SCHOOL

A social geography of out of school childcare

Fiona Smith and John Barker

The rapid growth of out of school care

In this chapter we explore the social geography of an increasingly significant children's environment, the out of school club. Defined as providing childcare to groups of British children aged 5–12 in settings offering creative play or learning opportunities, the services discussed below are provided by adult childcarers, usually called playworkers, who supervise children and organise activities after school, before school and during the school holidays. The service is primarily provided to cater for the children of employed parent(s), although a minority of children chose to attend clubs to avail themselves of the social activities on offer. Though out of school clubs are managed by a variety of organisations and individuals, in a diverse range of settings (see Smith and Barker 1998, 1999c for detailed profiles of provision), the majority of new clubs are provided in schools by voluntary management committees.

Since the beginning of the 1990s, the rapid growth in out of school care has had a profound impact on the social and cultural landscape of childhood in Britain. The number of out of school clubs expanded exponentially throughout the decade as a response to economic restructuring and the feminisation of the labour market (Smith 1996). Recognising the importance of out of school care to economic growth, the Conservative Government launched the 'Out of the School Childcare Initiative' in April 1993. As part of the initiative, the Department for Education and Employment invested nearly £60 million to expand the number of out of school places around the country by more than 50,000. The explicit aim of the initiative was to enable mothers to (re)enter employment by improving both the quality and quantity of out of school childcare available in the local area.

Shortly after coming to office in 1997, the Labour Government revealed its commitment to the continued development of the service by outlining its proposed unprecedented expansion of out of school childcare. The new government pledged to invest £300 million in out of school provision over a five-year period to create up to 30,000 new out of school clubs, with places for up to one million children. As the then Social Security Secretary, Harriet Harman, stated: it was the aim of the government to provide 'out of school childcare for every community in Britain'. Out of school care was also highlighted as a key part of other major new government initiatives including 'Welfare to Work', and the 'National Childcare Strategy'. As a direct result of these policy initiatives, the number of out of school clubs grew from only 350 at the beginning of the decade to over 4000 at the end, representing an increase in provision of more than 1000 per cent (Smith and Barker 1999c). The resultant expansion of clubs is thus clearly impacting upon the lives of a growing number of children, who are spending an increasing amount of time within an out of school club environment.

The institutionalisation of childhood in out of school clubs

The investment by both the Conservative and Labour Governments has obviously had a profound impact on the expansion of the out of school service and thus on the spatial and institutional framework of childhood in Britain. The growing number of children in out of school care is playing an important role in the institutionalisation of childhood, which 'refers to a process by which organised arrangements, chiefly the school system, influence children's lives and organise their days' (Frones 1994: 150). The expansion of out of school care, and the subsequent rise in the number of professionals and semi-professionals working for services, is a clear example of the shift in functions associated with childhood which were once performed in the domain of the family and are now increasingly performed in the domain of the childcare institution. The time spent in the institutional settings of the school and the out of school club is increasing for a growing number of British children. In fact, for children in full-time out of school care, the vast majority of their waking hours are spent in an institutionalised setting. Childhood for these children is extensively subjected to spatial and temporal control by teachers and childcare workers (Nasman 1994). The institutionalised environment of the out of school club therefore represents another place in the social and cultural landscape of childhood where adults try to shape children's use of space and time.

The compartmentalisation of children into institutionalised childcare settings, physically separate from the spaces of adulthood, is exacerbating one of the most important features of the modern conceptualisation of childhood: spatial separateness. This separateness is expressed through a whole range of cultural practices, which emphasise that children are distinct from adults. As a consequence of the way we conceptualise children, they are sent to school and to out of school clubs in order to be cared for, protected and to learn and acquire the skills associated with adulthood. As James (1990: 282) argues, 'certain activities and spaces come to symbolise childhood'. Out of school clubs provide a pertinent example of such 'spaces'. They thus provide geographers interested in children's environments with significant new sites in which to undertake research.

A social geography of childhood: the case of the out of school club

The newly emerging social geographies of children and childhood draw upon the work of a small but growing number of geographers interested in children's experiences and use of social space (Aitkin 1994; Philo 1997; Sibley 1995). Central to the development of such work is the notion that children are active agents, acting in a social world that has been structured largely by and for adults. The 'out of school club' provides a pertinent example of such a social space of childhood, as its development has been, and is being, instigated, defined and controlled by adult users (parents, employers, local authorities and so on). As Sibley (1995: 130) argues: 'In order to understand how children experience the [out of school club], we need a perspective which focuses on power relations, the way power is expressed in [adult/child] interactions and played out in the spaces of the [out of school club]'. In this chapter we will attempt to provide such a focus by exploring the ways in which children experience, (re)interpret and (re)negotiate the social space of the out of school club. By drawing on a variety of research methods carried out with approximately 500 children attending out of school clubs in England and Wales[1] (Smith and Barker 1999a, 1999b), we aim to explore how children experience this increasingly important environment in 'both its oppressive and liberating aspects' (Sibley 1995). In so doing, we aim to illustrate how the concept of space in these places is inherently contested between the children who use it and the adults who attempt to control it. We explore the ways in which age and gender impact upon children's ability to control the use of space within clubs. Other social variables, though also clearly important, are not explicitly considered in this chapter due to the constraints of space. Moreover, we aim to highlight how the wider institutional environment of the school in which the club is located impacts upon this process. In the rest of this chapter we explore the social geography of the out of school club by drawing on some of the key themes that children taking part in the research have identified as fundamental to their experiences of this increasingly significant social space of childhood.

A place to play: the struggle for control

The vast majority of children taking part in the research conceptualised the out of school club as a place to play after school and during the school holidays. Though they acknowledged its role in providing childcare for them and clearly understood they were attending whilst their parents were at work, the opportunity to play with friends and to try different play activities were central to the ways in which they experienced the social space of the club, in which they spent a considerable amount of time:

LAURA: You get to play games and you get to play with some people that are your friends.

(Girl, aged 10)[2]

In most of the clubs visited by the researchers, the play activities on offer were defined and planned by adult staff. In the minority of clubs which consulted with children, their suggestions were used by playworkers as a pool of ideas from which the staff short-listed and purchased games according to adult defined criteria, such as suitability for the age groups involved, financial cost, maintenance and sometimes educational value. With a few exceptions, the children were absent from making decisions about the types of play activities on offer in their clubs. The social space of the club was therefore planned and organised by adult staff, an example of the inequitable distribution of power between children and adults inherent to the vast majority of the clubs taking part in the research.

Despite this, the research highlighted a plethora of ways in which children contested both the activities provided by adult staff and also adults' control over the way space was used and structured. Far from being the passive recipients of adult-defined care, therefore, children were actively engaged in the (re)definition of activities. One club, for example, had a weekly theme to provide an educational focus to the games and activities provided. The staff decided a very large cardboard box was to be decorated as a Trojan Horse for the 'Ancient Greece' theme. However, instead of using the box as a Trojan Horse, the children created a competition to see how long each person could spend in the box. The role of the staff changed from one of educating the children about Ancient Greece, as had been planned, to one of ensuring that children kept an orderly queue outside the box. As one child said:

RYAN: Yesterday we were playing in a box . . . which we thought was quite fun.

(Boy, aged 9)

The creation of dens also proved to be popular with children attending the out of school clubs taking part in the research. By moving furniture from its designated position, utilising dressing-up clothes and requesting sheets that would normally be used for 'arts and crafts' sessions, it was common for children to convert parts of the club into dens where the presence of adults was strictly prohibited. By creating structures within the club which were physically too small for adults to inhabit, children were able to exert some ownership of part of the spatial environment.

These examples highlight just two of the ways in which children successfully (re)claim some control over what goes on in out of school clubs. It should, however, be acknowledged that the ability of children to contest and redefine their play space varies according to the age and gender of the children concerned. Younger children's attempts (for example, to make dens or create imaginary racing tracks for cars) were interpreted by adult staff through traditional constructions of childhood as innocent and playful (see Chapter 1). Perhaps somewhat ironically, these stereotypical notions of childhood, institutionalised in clubs by the practices of adult staff, served to enable rather than constrain younger children's attempts to redefine the social space of the club.

Older children's attempts to, for example, create an older boys' room in one club and claim 'ownership' of the television room in another, were regarded without exception as problematic. As such, they were treated as a hostile attempt to subvert the workings of the club by playworkers. Staff were thus far more likely to intervene and restrict older children's attempts to (re)claim part of the space of the club as their own, viewing them as 'young anarchists' rather than as 'playful, inquisitive young children'. Whereas the older children themselves were acutely aware of the significance of age on the way different children experienced and felt about the out of school clubs they attended, adults' reactions were embedded within the dominant 'little angel/little devil' discourse, juxtaposing and polarising the qualities of childhood according to age.

BILL: Sometimes it's a mare,[3] because most of the things are for young children. This [club], it's mostly for younger children.

(Boy, aged 9)

CHRIS: They [younger children] think it's perfect. Once a girl was crying because she didn't want to leave.

(Boy, aged 10)

Gender was also a significant variable impacting upon the way the use of space in the club was contested by different groups of children. Boys and girls challenged adult-defined play activities within the clubs in different ways and vied for control of different parts of the physical environment. Boys generally attempted to gain control of the outdoor space, wanting to play football and other physically active games. Girls struggled for control of parts of the indoor environment where they could play in private. In particular, girls attempted to lay claim to parts of the club where they could talk about and sing Spice Girls songs without interruption from either adults or boys.

The ability of boys and girls to gain control of different parts of the club was ultimately in the hands of the adults, who retained the balance of power in all of our case studies. Once again, playworkers drew upon institutionally embedded conceptualisations of acceptable play. They were more likely to view girls' requests positively, since their activities (for example, emulating their favourite pop group through role play) were seen as elements of 'successful' play. Adults largely interpreted boys' attempts to play more football and other playground games as problematic. This was primarily due to the fact that boys prevented girls from joining in by ridiculing them and making them feel they could not play:

YVONNE: . . . and when I used to play football, they used to make fun of me because I can't really play.

(Girl, aged 7)

Boys' attempts to control the football pitch or playground were likely to be challenged by staff since they clashed with adult-defined 'equal opportunities' policies. Many boys understood this process in terms of gender relations, arguing that female staff were unable to relate to boys in the club:

LEO: They treat girls differently, and they treat boys like they are things that don't belong here.

(Boy, aged 10)

The struggle for control over the physical environment of the out of school club was clearly evident in all the case studies we visited whilst undertaking this research. Thus the structure of the play space, although initially designed by adult playworkers, was continually challenged and (re)structured by the children. The spatial environment of the out of school club became a site of contestation, creating a social geography of out of school childcare layered with inequitable power relations between different groups of children and between children and adults.

Playing 'out of school', in school

For children attending out of school clubs run in schools, the ability to challenge the use of space in clubs was further limited by the wider institutionalised environment of the school. Many clubs, for example, were held in school halls that contained gym equipment. With one exception, the gym equipment was outside the boundaries of the play space designated to the out of school club. The issue was confusing to children because of the physical location of the equipment within what they understood to be the geographical boundary of club.

ALEX: I'd like to get the apparatus out . . . and play on the rope ladders and stuff.

(Boy, aged 9)

SIMON: You're not allowed to go on the PE apparatus.

(Boy, aged 8)

CONNOR: You're not allowed behind the PE equipment.

(Boy, aged 6)

Playworkers strictly reinforced the boundaries defined by the school when children attempted to contest the notion that the gym equipment was 'out of bounds', for example, by trying to play on benches and mats. The playworkers understood the children's wishes, but explained to the researchers the legal, financial and insurance implications of using such equipment, and the potential difficulties of setting up, putting away and maintaining the equipment at a level demanded by the school. Playworkers effectively policed the boundaries enforced by the school environment. The ability of children to challenge, contest and (re)shape their club was therefore limited by the constraints imposed by the wider institutional environment of the school.

This tension was also illustrated repeatedly by the design and decoration of the built environment of the out of school club. Many children thought that the appearance of clubs held in schools was drab and uninspiring, needing colour and pictures on the walls to (re)establish its role as a play space rather than as a school space.

NANCY: I think the building should be more colourful . . .

(Girl, aged 9)

KEVIN: Yeah . . . I think it should be more nice.

(Boy, aged 6)

NANCY: You could paint it with colours . . .

(Girl, aged 9)

KEVIN: . . . you could make lots of pictures and put them up.

(Boy, aged 6)

Most clubs, however, were prohibited from displaying such pictures and posters. Clubs were also often prevented from adapting the built environment in any way. In one, children complained that the lack of comfort was a problem:

STEPHEN: The floor is very hard, so when you want to read you have to sit on a hard place.

(Boy, aged 6)

NANCY: I think we should have bean bags to sit on, for every person.

(Girl, aged 9)

The use of bean bags was not permitted, primarily due to the lack of storage space available to the out of school club. Its facilities and equipment had to be packed up and put away at the end of every session. Thus the out of school club was transient, only visible during opening hours. At all other times, the space was devoid of any signs of the club. This was because the school hall, in which the club ran, was used by other groups at different points in the day and at weekends. The school, under pressure to generate income from their premises, did not feel they were in a position to allow the children to alter the décor of the space used by the out of school club. The children's desires to decorate their club was therefore constrained by the other institutional demands put on the space. From the point of view of the majority of the children, this left the club without a distinct identity, leaving it with a 'school like' ambience that they felt was inappropriate for a 'play' environment.

One club had successfully negotiated with the school to allow children to decorate the space in which the club was held. Based in the school canteen, the children filled the walls with drawings and pictures created whilst at the club. The children were clear that their pictures and drawings enhanced the look of the canteen, provided the club with its own identity and gave them a sense of ownership of the environment. The ability of children to decorate this environment was facilitated by financial bargaining between the school and the out of school club. Children were given the opportunity to redesign what they saw as their 'play environment' because the club agreed to pay for the decoration and upkeep of the canteen. It is significant that children's control over the visual appearance of this space was only realised through negotiation between adults, a process that excluded them.

Furthermore, although out of school care is provided in schools outside the hours of the formal school day, clubs taking part in the research were invariably subject to many of the institutional practices of the schools in which they were located. Rules and regulations were imposed by playworkers to ensure that children behaved in a way deemed appropriate by the school. In after school clubs in particular, children's activities were institutionalised by unofficial, but regularly imposed, 'supervision' by teachers and other school staff. This supervision took both covert and overt forms. In many cases, children were told to play quietly at the club, for fear of disturbing teachers at work in other parts of the building. In other cases, school teachers routinely carried out their own after school duties within earshot or sight of the activities of the club. For many playworkers, this was seen as a way in which teachers 'kept an eye on' the children's activities. It was common for clubs taking part in the research to articulate that they felt they were under surveillance from the school in which they were located. At two of the clubs visited, this surveillance took on a more active form of social control. School teachers entered these clubs and took charge at times when they felt the playworkers were not controlling the children's behaviour in an appropriate way. Thus, despite the fact that teachers had no official jurisdiction in the out of school clubs we visited, they clearly considered themselves to be more powerful than the playworkers who were officially in charge. Teachers, therefore, represented a further layering of power through which the social space of the out of school club was challenged, contested and ultimately (re)structured.

The 'right' time to play?

The space of the school hall changed its social significance for children at different times in the day. Its use as a site for assembly in the morning, PE lessons throughout the day, lunch-time activities and play after school highlights the fluidity of meaning attached to this space. One of the ways in which children negotiated the temporal transformation of the 'educational' space of the school hall into the 'play' space of the out of school club, was by engaging in activities explicitly associated with the school playground and school playtime. Whilst playworkers attempted to interest children in activities that were not offered in school, children themselves (re)structured the out of school club as a school 'playtime' environment. Thus children engaged in playground style games such as football, 'bulldog' and 'it',[4] (re)constructing the out of school environment on their own terms. Education, play, time and space became fluid notions for the children as the power to define their use of space was institutionalised by both the out of school club and the wider environment of the school in which the club was located.

Though most clubs held in schools had daily access to the school hall or canteen, the actual physical location of the clubs visited was not fixed. Other areas of the school, such as the library, computer room and playground, were also sometimes available. Many clubs were forced to move site on a regular basis due to demands for space from more powerful adult users. In a number of cases, for example, parents' evenings forced the out of school club we were visiting to vacate its usual site. Children's use of space after school was therefore shaped by the wider commitments and timetables of the school. Children, arguably the main users of this space, were not consulted in this process and were thus confined to the parts of the school not in demand from the more influential (and in the eyes of those who ultimately controlled the space, more important) adult users. Thus, while children could sometimes successfully contest the use of space within the out of school club, they had little power to define where the club was held or to widen its boundaries within the wider institutionalised environment of the school.

Conclusion

In this chapter we have begun to create a 'genuinely social geography' of the out of school club, by exploring how the concept of space is contested between the children who use it and the adults who attempt to control it. While the social distance between children and adults has clear implications for the sorts of power relations inherent in clubs, it is clear that far from being passive recipients of out of school care, children are active agents, challenging adult notions of appropriate behaviour and the use of space. The examples provided in this chapter illustrate some of the ways in which children are actively engaged in the (re)construction of their own out of school environments, a process they clearly find liberating.

The research undertaken for this chapter in a number of out of school clubs throughout England and Wales, has highlighted that gender and age are key social variables impacting upon children's ability to successfully contest the way space is structured and used. Boys and girls of different ages attempt to redefine the social space of their clubs in a variety of ways. Moreover, though these practices themselves originate from children's actions, they are largely mediated and interpreted by more powerful adult staff, who draw upon institutionalised definitions of 'acceptable play' and 'childhood' to intervene and resolve children's conflicts from their own, adult, standpoint. It is evident that adult staff and the wider institutionalised environment of the school impose constraints on the ability of children to use the space of the out of school club freely. The physical boundaries associated with where children can or cannot go, are actively policed by playworkers, under the surveillance of teachers. The social geography of the out of school club is therefore subject to a layering of power between different groups of adults and different groups of children, as well as between adults and children. Children's experiences of out of school care are subject to wider institutional processes over which they have little control. Children experience the fact that they are unable to use certain pieces of equipment, or to play in particular places that have been accessible to them earlier in the day, as oppressive and confusing. The shifting boundaries between school and the out of school club, mean that the concepts of 'school time' and 'out of school time' are often blurred and perplexing for the children.

Children's use of space in out of school clubs is therefore subject to both adult/child interactions and to the wider environment within which the club is located. If children are to spend an increasing amount of time 'out of school'/'in school', as indicated by the current growth of out of school services, we suggest that it is imperative that service providers ensure their experience is as 'liberating' as possible. The research undertaken for this chapter suggests that children are happiest when they have some control over the way their clubs develop. However, as this process necessitates the (re)conceptualisation of children as competent social actors by service providers and schools, it is questionable how many children will be actively engaged in the future development of out of school care.

Notes

1   This research is funded by the Economic and Social Research Council (project Number L129251050) as part of the Children 5–16: Growing into the 21st Century Research Programme.
2   The names of the children have been changed as agreed at the time of obtaining informed consent from them.
3   'Mare' is slang for 'nightmare'.
4   'Bulldog' and 'it' are common playground games in the UK, where children have to run after and catch one another.

References

Aitken, S. (1994) *Children's Geographies*, Washington, DC: Association of American Geographers.
Frones, I. (1994) 'Dimensions of childhood', in J. Qvortrup, M. Bardy, S. Sgritta and H. Wintersberger (eds) *Childhood Matters: Social Theory, Practice and Politics*, Aldershot: Avebury.
James, S. (1990) 'Is there a "place" for children in geography', *Area* 22, 3: 278–83.
Nasman, E. (1994) 'Individualisation and institutionalisation of childhood in today's Europe', in J. Qvortrup, M. Bardy, S. Sgritta and H. Wintersberger (eds) *Childhood Matters: Social Theory, Practice and Politics*, Aldershot: Avebury.
Philo, C. (1997) 'War and peace in the social geography of children', paper delivered at a meeting of the ESRC Children 5–16 Research Programme, University of Keele, March.

Sibley, D. (1995) 'Families and domestic routines: constructing the boundaries of child-hood', in S. Pile and N. Thrift (eds) *Mapping the Subject: Geographies of Cultural Transformation*, London: Routledge.

Smith, F. (1996) 'The geography of out of school care', unpublished Ph.D. thesis, University of Reading.

Smith, F. and Barker, J. (1998) *Profile of Provision: The Expansion of Out of School Care*, London: Kids' Clubs Network.

—— (1999a) 'Learning to listen: involving children in the development of out of school care', *Youth and Policy* 63: 38–46.

—— (1999b) 'From "Ninja Turtles" to the "Spice Girls": Children's participation in the development of out of school play environments', *Built Environment* 25,1: 35–43.

—— (1999c) *The Childcare Revolution: Facts and Figures*, London: Kids' Clubs Network.

15

NATURE'S DANGERS, NATURE'S PLEASURES

Urban children and the natural world

Lily Kong

Introduction

Playing

Mummy, I want to go to the park to play!

> (Daniel, 8-year-old, interrupting an interview)

Don't climb the tree – you'll fall down and hurt yourself.

> (Anxious mother, overheard during participant observation)

Learning

Mummy, what is that insect? Why is it such a funny colour? Is it dead? It hasn't moved for a long time.

> (Rachel, overheard during participant observation)

I think the children enjoy doing gardening. They sort of help pull out weeds . . . getting them interested, teach them why these plants have to be pulled out, why those ones you leave and those ones you don't.

> (Melanie, grandmother of three, focus group discussion)

Caring

Mum let me buy some seeds with my pocket money, and I grew some flowers in pots. The first time, the periwinkles grew and I was so happy.

> (Emma, 8-year-old).

In the past decade and a half, growing attention has been paid to the notion of nature as a social and cultural construct (see, for example, Evernden 1992; Fitzsimmons 1989) as well as the need to understand at an everyday level, human relationships with and experience of nature (see, for example, Burgess, Harrison and Limb 1988; Burgess, Limb and Harrison 1988; Kong *et al.* 1997, 1999). Clearly, different social groups construct, access and experience nature in different ways, and there is a need for more detailed and nuanced understandings of such different constructions and experiences. Though some work has been done on how women in different cultural contexts experience nature (see, for example, Burgess, Harrison and Limb 1988; Burgess, Limb and Harrison 1988; Kong *et al.* 1997), far more work needs to be done, taking into account issues of race, class and age, for example. In this chapter, my intention is to focus specifically on children[1] in a highly urbanised setting and explore their constructions and experiences of nature, using Singapore as my case study. I have chosen to focus on an Asian and highly urban context, given what I see to be an urgent need to understand urban dwellers' relationships with nature, particularly since the world, and Asia in particular, is urbanising rapidly. Given preliminary evidence about the importance of childhood experiences in cultivating a love for nature (Kong *et al.* 1999), I have given specific attention in this chapter to children, with a view to understanding some of the possible long-term implications of childhood experiences of nature. In what follows, I will use the three experiential and conceptual categories that I began the chapter with to frame my discussion, given that my empirical work in Singapore suggests that these three experiential realms form the broad contours by which children conceive of and experience nature.

Research context

Geographical research on children may be classified broadly in terms of exploration of the environmental behaviour of children and the development of children's knowledge of the geographic environment (Hart 1984). The former, which Hart termed the 'geography of children', is concerned largely with children's spatial behaviour, and draws heavily on work in psychology. Among the research issues are the life spaces of children, including, for example, the spatial ranges and restrictions of children, and the importance of the home-base in child development. The second area of research, which Hart termed 'children's geographies', is focused mainly on children's understanding of spatial location and phenomena and their spatial awareness. This characterises most of the early geographical research on children, beginning in the 1970s, and was very much influenced by developments in behavioural geography, with its concepts of environmental perception and mental maps. This was initiated mainly by Blaut and Stea's (1971) work, which spawned a significant body of research (Catling 1979;

Dijkink and Elbers 1981; Downs, Liben and Daggs 1988; Jackson and Johnston 1974; Matthews 1984, 1986, 1992; Spencer, Blades and Morsley 1989).

In the 1980s and 1990s, research on young people and their environments, though still somewhat limited, became more theoretically sophisticated, inter-rogating the symbolic meanings of places and the negotiation and contestation of meanings of places for young people (see, for example, Kong and Ng 1996; Rutheiser 1993; Sibley 1995; Valentine, 1996). Such investigation of the symbolic meanings of places serves as the springboard to my empirical analysis of how children in urban Singapore construct and experience 'nature' and the symbolic meanings and values that they invest in nature. Before embarking on the empir-ical analysis, however, I will first paint in some background on Singapore and the methodology used.

Singapore: the triumph of planning or the crisis of nature?

Singapore is a city-state, and as the nomenclature implies, is characterised by its highly urbanised setting. This was, however, not always the case. Clearance of natural areas began in the late nineteenth and early twentieth century, and momentum gathered particularly in the post-1960s after independence from British colonial rule. When the newly constituted government of Singapore achieved internal self-government in 1959, it was confronted with a plethora of problems, particularly rapid population growth, housing shortages, high unemployment and poor infrastructure. As part of the national efforts to address these problems, various economic and social programmes were initiated, which necessitated an immense degree of land-use planning and land and building development. Such planning and development often entailed clearance of natural areas, from forests and ridges to swamps and coral-fringed coasts, as well as damming of rivers for reservoirs. The result was that the proportion of Singapore covered by forests decreased from 6.5 per cent in 1960 to 4.4 per cent in 1996, while the proportion covered by mangroves dropped from 7.9 per cent to 2.4 per cent in the same period. Correspondingly, the proportion of built-up area almost doubled from 27.9 per cent in 1960 to 49.3 per cent in 1996 (Ministry of Information and the Arts, 1996; Wong 1989: 774).

Just as Singapore's natural areas have faced clearance over the years, other forms of nature have been 'constructed' with the aim of satisfying various human needs. These take a variety of forms, for example parks that have been specially designed, constructed and planted with vegetation; trees and shrubs particularly selected for roadsides, road dividers, car parks and other such open spaces; creepers trained onto walls and overhead pedestrian bridges and so on. With this 'construction', Singapore's landscape transformation from dense tropical forests to an equally dense built-up environment has meant that natural areas continue to be destroyed while various policies and actions have been simultaneously introduced to green the city. The result is that the form of nature which Singaporeans have become familiar with is managed messicol vegetation[2] which was deliberately planted to provide some balance in an increasingly urban environment. For the average Singaporean, there is little contact within Singapore with naturally occurring unmanaged greenery and wildlife, and it is in this context that most young people have grown up in the past three decades.

This chapter draws its material from interviews with twenty children about interaction with and feelings about nature, as well as focus group discussions with youths (tertiary students in their late teens to early twenties from universities and polytechnics), mid-teen students (all drawn from one secondary school), users of a neighbourhood park in Clementi, a public housing estate in Singapore, and an all-woman group (all drawn from a community church). The interviews with children were informal and often wide-ranging, given the short attention span of children. They were often conducted over a meal. The children were accessed through a neighbourhood school and local church, and included an equal number of boys and girls. Each focus group, in turn, comprised between four and twelve persons, including both males and females (except for the last group) and each group met once for an hour and a half to two hours. The tertiary students were recruited through student organisations in three tertiary institutions. The teenage students were recruited with the help of a teacher from a school who sought volunteers from her classes. The all-woman group was recruited through a parachurch group leader. In this focus group mothers spoke about their children's experience of nature and recalled their own childhood experiences.[3] I wished to pay attention to the importance of the interplay of discourses on nature by both the young and adult populations because I felt that young people's access to and experience of nature are mainly negotiated within families. I also wish to emphasise that young people's experiences are not monolithic and that some differences (but also similarities) were evident in the ways in which young girls and boys articulated their experiences with nature. Such divergences, however, are not my main concern here. Given the paucity of research on children and their construction and experience of nature, I am more concerned to paint on a broad canvas in this instance and leave more detailed analysis to subsequent work.

Constructions and experience of nature

Playing

The primary theme around which discussions and interviews revolved was the notion of 'play' and the role of nature in children's recreation. Two views were apparent among both children and adults. One underscored the pleasures of play in nature; the other accentuated the dangers.

One of the key ways in which mothers (especially) spoke about nature was in terms of how their children experienced (or should experience) nature. In particular, they focused on the opportunities for recreation that nature afforded to their children. Such opportunities were thought to be healthy and necessary for children to grow up into robust, well adjusted adults. As two mothers in their late thirties, Grace and Agnes, pointed out, children needed plenty of space to run about, to expend their energy, so they would bring their children to the public parks and gardens for those precise reasons. But beyond the openness of such spaces, nature also offered other recreational opportunities to children: to feed the fishes, climb trees, pick shells and build sand castles. Such activities, they recognised, afforded opportunities for children to grow and develop in the midst of play, as expressed in the context of Britain:

> the environmental experiences available for children, especially those in middle childhood when they encounter the world for themselves, away from grownups and parental control, are crucial in the development of sane, healthy adults with an appreciation for nature. The needs of children to be able to explore in 'safe danger' as one mother put it, . . . related directly to the desire to instil a sense of respect and value for the natural world.
>
> (Burgess n.d.: 24)

As the mothers (particularly) in the focus groups emphasised, nature could provide abundant pleasures if they could 'let go' and allow their children to take physical risks, which they acknowledged as an important part of growing up.

Although adults affirmed the need for 'healthy fun' in nature, drawing particularly from their own experiences of childhood fun, their conception of nature as a safe and healthy recreational space for children is not unmediated. Many parents admitted to their tendency to worry. Grace, for example, recognised that she had a 'phobia about the children falling down and things like that'. Indeed, as observed in a recent report in Singapore's national English daily,

> Singapore children don't seem to run around playing in green fields any more. It used to be that one couldn't go past a field on the island without seeing bands of children racing about in it in great glee. Indeed, one couldn't even go past a stony and uneven patch of land without seeing happy kids using it for some activity or other. Nowadays, there are a great many more well-trimmed and levelled green fields available

for kids to play in, but one hardly sees any children using them, let alone having fun in them – not even in fields near big Housing Board estates.

(*The Straits Times*, 16 June 1999)

This, Melanie argues, is because

children now are too controlled. You see, in our days, children would like to climb trees and pluck cherries . . . whatever there is they can do. Now, parents have one child, two children. They are so scared that they will fall down and break their heads. I mean, I have two grandsons. Their parents are so worried. They do not have the experience of being up on the tree and they are so scared. They would rather let them sit at the computer rather than let them go out and enjoy the trees and the flowers. I think it's the parents now.

Thus, while Singaporean adults recognise the value of recreation in nature, at the same time, parents with few children or even one child tend to adopt overly protective attitudes which prevent children from exploring nature's pleasures, emphasising instead nature's dangers. This may be an unintended consequence of Singapore's highly successful population policy of the 1960s, 1970s and a large part of the 1980s, which emphasised small families (Perry, Kong and Yeoh 1997).

This simultaneous predilection towards and disinclination towards recreation in nature is also evident among the children I spoke with, and often in more pronounced ways too. For example, Serene, a 12-year-old, expressed a dislike for the heat of the sun and fear of water, preferring instead computer games, the television and karaoke. Twelve-year-old Robert admitted that he went to the park only because his mother insisted that he go for walks with her and the family. In response to a comment about the beauty of flowers and the wonders of the smells of nature, he said: 'Nothing. I can't smell anything. I go to the gardens or whatever, but I can't smell anything.' Amidst laughter from the rest of the discussion group which he had attended with his uncle, and general agreement that 'it's the age' that precludes him from enjoying what other group members enjoyed, it became evident that young people like Robert preferred a whole array of other activities for entertainment, ranging from music to shopping to sport. Robert illustrated this when he emphasised the point that the enjoyment derived from contact with nature that the other group members enjoyed was, in his mind, 'a waste of time'. Some of the children I interviewed also generally expressed the view that, when bored and thinking about what places to visit and what things to do, their tendency was not to think of activities associated with nature. When thoughts about the natural world did surface in their minds, they admitted, it was often in the context of school work, for example their geography and science lessons, during which nature was more about conceptual issues and scientific processes than everyday environments of potential fun and enjoyment. Nine-year-old Timothy's comments sum up the view best:

> When I don't have to do my school work, I like to go skateboarding or play football. I also like the computer games. My cousin, Robin, has a lot of new games . . . No, I don't think of going to the park or to see birds but if I have to collect some things like leaves for science class, then I have to go.

There were, however, some who had particular predilections for recreation in nature, although often they also spoke in the same breath of resistances from their parents towards such activities. Eleven-year-old Mary, for example, expressed frustration that her mother frequently stopped her going cycling in the park and would 'freak out' if she knew that her daughter climbed trees to pluck fruits with her cousins. Twelve-year-old Mark lamented that his parents would not allow him to go with his friends and their families to Pulau Ubin (one of the few islands making up the Republic of Singapore which remains relatively natural) and Sungei Buloh Bird Sanctuary because of the fear of unknown insects and animals. Some children's disinclination and many parents' overprotection combine to produce the empty fields, described earlier, which are symptomatic of children's lack of contact with nature in urban Singapore. This occurs despite the recognition, at least on the part of parents, of the value of recreation in nature. A slippage therefore exists between discourse on nature as recreational pleasure for children and actual physical engagement.

Learning

Children appear to have a curiosity for what they see in nature, making it a living classroom. Nine-year-old Audrey described with obvious joy trips to the Bukit Timah Nature Reserve with her zoologist father and older brother:

> Dad shows us some birds, and they're not just all 'birds' like my friends think they are. They are different, sing differently, and you can spot them if you look carefully. Then, there are the lizards and he'll turn the stones and look for some insects underneath. I go to school and can tell Miss Lim [science teacher] what Dad showed me, and she won't know from the books what I'm talking about!

Grandmother Melanie similarly takes her grandchildren for walks along a nearby green corridor, and points out plants to them, naming them and explaining the processes of flowering and fruiting, and other natural processes. Evoking her role as a science teacher before she retired, Melanie speaks animatedly about nurturing children's knowledge of and interest in nature not only from information imparted, but from curiosity and love developed through doing. She therefore gets her grandchildren interested in her garden and involved in her gardening, as one of the opening quotations of this chapter exemplifies. In turn, her grandson, Joe, testified to the fun he had in 'grandma's garden', because he felt 'good about learning what were weeds and therefore had to be removed to ensure the survival of other plants' and he had learnt enough to be 'in control of the patch'. Indeed, many mothers in our discussion groups saw their role as awakening their children to the wonders and workings of nature. Sally, for example, a 32-year-old mother, spoke approvingly of her sister-in-law's garden in which there are all kinds of fruit trees; neighbours who pass by would stop to show their children what a mango tree or tomato plant looks like. 'If I had a garden, I would do just that. It's great for children to learn!' Her daughter, 12-year-old Susan, was even more eloquent in expressing the uniqueness of learning from nature itself:

When I learn about how fruits form in the textbook, there are descriptions and maybe diagrams, but if you actually see it happening and growing, you actually understand what you're reading. All of a sudden, it becomes real.

Nature's potential as a living classroom, however, is not always realised, because adults themselves do not have the knowledge to share with their children. Kathleen and Meilin, women from two separate discussion groups, express frustration at their inability to tell their nieces and nephews the names of trees when they asked them. However, if they did not have the knowledge themselves, they often expressed the view that more could be done in schools and other public places to engage children with their living classroom. Carrie, a mother of two, for example, insisted that parks in housing estates should have labels against trees and other plants so that children would know what they were looking at. Grace expressed the view that children, when learning science, should be taken out of their classrooms so that they can

feel what's this leaf like, what's the shape of this leaf, why is it different from this plant, in what ways is it different. I mean, I miss knowing all these also when I was in school and I think that it's very sad because you know, I only learnt from the book, and it's not real to me. It doesn't come alive . . . We should start to arouse the kind of interest and curiosity in children.

When curiosity is aroused, however, and questions cannot be answered, children quickly appear to lose interest and become bored. Hence, Rachel, quoted at the beginning of this chapter, was observed to run off to play catching with her siblings and friends when her host of questions about an insect she had observed could not be answered.

Several of the mothers, however, recognised the difficulties and lack of opportunities to harness nature as a living classroom in urban Singapore. As Ellen lamented, schools are now found in very built-up areas and are often not near natural areas. Children therefore have few opportunities to learn in the huge science laboratory that is the natural world. This is in contrast to her own childhood, when she learnt about the barks of trees by running into the school garden and doing bark shading and colouring.

In sum, there is no lack of recognition of the value of nature in the learning process. Children can enjoy and be fascinated by the processes and products of nature if they are pointed out to them, and if their curiosity is satisfied with answers. Adults similarly remember what they learnt in the 'living classroom' and are keen to share their lessons with their children. This potential, however, has been eroded with the continued urbanisation of the island. Yet, even within the highly built-up surroundings, it is possible to harness nature's potential as a classroom more fully, drawing on small pockets of blue and green in neighbourhood parks, introducing more elements of nature in school compounds, and using the limited opportunities of exposure to nature more frequently and creatively. Providing access to nature is, however, but one aspect to be tackled. As apparent from earlier evidence, children may actually be bored with nature if they understand nature's processes only as textbook knowledge and do not relate them to real processes of growth and decay in the garden. Children may also become bored with nature when their myriad questions are repeatedly not satisfied.

Caring

In one of the opening quotations I used in this chapter, I drew attention to Emma's joy, derived from nurturing the growth of flowers from seeds. Such sentiments are expressed by some of the other children I talked with as well. For example, Joe and Mabel (both 11-year-olds), talked about the Nature Club at school and the caring and tending of pets and plants as part of the activities. Joe talked enthusiastically about recess periods in between classes and the joy of running to feed the hamsters and rabbits, and stroking and playing with them, and especially looking after 'Hammy the hamster' when he injured himself. Mabel, too, expressed a sense of triumph at seeing her beans sprout as she tended them daily. Yet, these children were a minority in expressing these joys of caring. Most had little interest in plants and animals, and indeed, several children living in highrise settings not only did not express care and love for nature, but feared touching or going near even domestic animals. Twelve-year-old June exemplified such children:

> I'm afraid of the dog because it might bite me, and they always like to come and jump at me. They're so noisy, always yelping away. They are so wiggly and furry. And cats also. In the coffee shop, they are always frightening me when they come to my legs or they brush their tails against me [she squirms at the thought].

These are children who have grown up with no contact with animals, because highrise living comes with its own set of constraints, including restrictions against pets such as dogs and cats (aquarium fish would be an exception since they can cause no nuisance to neighbours). Children's fear of animals, stemming from their lack of contact with any, underscores the importance of childhood contact with nature if there is to be an adult world comprising people with an affinity or at least sympathy for nature. This view is emphasised by 40-year-old Irene, brought up in a Malaysian *kampong* (village):

> I was born in Malaysia and grew up there. You are more exposed to nature there, so you appreciate the beauty, the nature. You're more in contact with nature . . . When we were small, we even went and caught spiders, swam in the river . . . We were exposed to that kind of environment, you see. So we grew to enjoy it . . . It's not inborn, this thing you cultivate. You learn to appreciate.

Her comments highlight the importance of childhood experiences in cultivating care and love for nature. For young Singaporeans today, the likelihood of such a *kampong* childhood is practically non-existent and the only other situation in which they are likely to have had a childhood characterised by close contact with nature is if they have spent some years overseas. Twenty-one-year-old Melinda exemplifies the case of someone who had spent three years of her childhood in the US and had therefore had the opportunity to go to national parks and camp grounds. She spoke of the wonders of nature and the simple joys of watching insects as they made their way from one point to the next, for example, pleasures she continues to seek and indulge in:

So I used to travel to those nature reserves, national parks. I just like to see animals, see how they behave. And even in Singapore, I think it's quite interesting and entertaining because in my garden, I look at how the birds . . . the way they behave, the way they eat the fruits from the trees. The way the ants crawl . . . it's very interesting.

Yet, as Grace, a mother of two from the all-woman group, suggested, because children today have little contact with and are not involved in tending and nurturing plants and animals, there is little love for nature. Indeed, she was confident that if a survey was done with children about what television programmes interested them, they would not mention nature documentaries at all. She felt strongly that without the opportunities for children to 'grow with nature', they would never 'have this love for it'. Similarly, Rita, a single woman in her mid-thirties lamented that parents nowadays did not bring their children to nature areas enough, resulting in a generation of young children who had little contact with and could not care for nature.

Once again, as with 'playing' and 'learning', 'caring' acts as a frame for understanding children's relationship with nature, but the lens is by no means one-dimensional. Clearly, care for nature can be nurtured, provided children have the opportunity for contact and interaction with nature. Without such opportunity, however, there can be little likelihood of any care and love for nature. Yet, opportunity need not come in the form of grand nature reserves and dramatic natural settings – the neighbourhood park and the school ecogarden are as important, if not more so, when adequately harnessed.

Conclusions

As research on children and their environment develop, it becomes clear that their needs and experiences may differ from that of the adult world, and that they deserve to be understood, first, as a separate group, and then, as a group which in itself is differentiated according to varied characteristics, such as gender, race and context (Matthews, Limb and Taylor 1999). In this chapter, I have sought to examine children's constructions and experiences of nature in the specific context of a highly urban setting. The data on which I have grounded my discussions in this chapter point to three types of involvement with nature. Specifically, through playing in, learning from and caring for nature, children are engaged socially, intellectually and emotionally.

This particular research illustrates how conditions predispose urban Singaporean children towards limited interest in and affinity for nature, to play, learn and care, even though such interest and affinity could have been quite easily encouraged. That is not to say that these Singaporean children's relationship with nature is uniformly 'flat'. Tensions exist in various guises: between a child's predilection for recreation in nature and an adult's fear of nature's dangers to children; between an adult's desire to engage in familial recreation in nature, and a child's preference for other forms of entertainment; between a child's desire to learn in nature and an adult's inability to respond to that desire; between children and adults' propensity to care for nature, and the lack of opportunity in urban Singapore to give full play to that emotion. These conditions are the result of a few factors: growing up in a highly urban environment in which contact with nature is limited; over-protective parents of two-children families who worry about the 'dangers' that their children are exposed to when playing in natural areas; and the abundance of other recreational and entertainment options. This poses a danger: that children become predisposed in their older years to adopt the rationality of the state when confronted with situations in which a development priority conflicts with the needs of wildlife and greenery. This is already borne out in my work with young adults (Kong, Yuen *et al.* 1999) who are quick to defend the need to remove natural areas, for example, for the construction of other 'more necessary' amenities. At the same time, if such natural areas are to remain, then they need to serve instrumental goals, defined in state terms variously as the provision of a salubrious and aesthetic environment, and indeed, to act as an economic resource (Kong and Yeoh 1996). In time to come, there is a danger that there may be a generation of adults with little concern for nature, and a lot fewer policy-makers and implementers with a genuine affinity for the natural world.

In adopting the position that I do, that environmental experiences available for children must allow them to 'encounter the [natural] world for themselves, away from grown ups and parental control' in order that they may develop into 'sane, healthy adults with an appreciation . . . sense of respect and value for the natural world' (Burgess n.d.: 24), I would recommend various policy and action programmes. Without suggesting that busloads of people be brought in to nature areas, it is nevertheless useful to consider introducing activities that will draw more people, including and perhaps especially children, to nature areas in Singapore. Some examples are those which involve the community in planting trees and tending the vegetation, whereby families or children could be involved. Educational visits for schoolchildren could be organised more frequently, involving learning through fun and games. At the same time, in the quest to bring cultural activities into the everyday lives of ordinary Singaporeans, cultural performances could be brought to parks, with organised school trips bringing children to culture and nature. Picnic spots could also be incorporated, although simultaneous educational programmes must also be launched to ensure that people do not vandalise or pollute the landscape with their litter. People should be kept informed of events, progress in their projects, and wildlife sightings, such as the arrival of migratory birds, which could prove exciting especially for children. In other words, children might be introduced to wholesome and fun activities in natural areas, ranging from parks to nature reserves, with some adult supervision and guidance. The types of places I recommend would be those everyday places which can be bound integrally to the lives of the community rather than specialist nature reserves for scientists and 'experts', although these more limited public access areas (because of the sensitivity of the flora and fauna to human disturbance) may periodically be introduced to young children as well.

What I hope to have achieved through this research is a beginning understanding of children's constructions and experiences of nature, within the context of a highly urban setting. My findings suggest some urgent need for educational and other policies to consider the social, intellectual and emotional development of children and the role that nature might play in that development. What is also evident is a need to reintroduce into Singapore children's lives more contact with the natural world, so that the empathy for nature may be inculcated from a young age.

Acknowledgements

Segments of this paper are drawn from Kong et al. (1997, 1999). Thanks are due to Blackwell Publishers and the Landscape Research Group Ltd for permission to reproduce some of the material.

Notes

1   I have adopted the category of 7- to 12-year-olds as my focus. These are children who I thought would be able to articulate their thoughts and experiences, but who had not crossed into the teenage years. They are all also attending primary school and therefore have a certain shared educational experience.
2   This refers to vegetation planted by humans and in a strict sense refers only to crops planted for harvest. Hill (1973: 31), however, extends the meaning to include plant communities which are deliberately planted and maintained for purposes such as aesthetic enjoyment and recreation.
3   Further details of the methodology employed may be found in Kong et al. (1999).

References

Blaut, J. and Stea, D. (1971) 'Studies of geographic learning', *Annals of the Institute of British Geographers* 61: 387–93.

Burgess, J. (n.d.) 'Rivers in the landscape: a cultural perspective', unpublished manuscript.

Burgess, J., Harrison, C.M. and Limb, M. (1988) 'People, parks and the urban green: a study of popular meanings and values for open spaces in the city', *Urban Studies* 25: 455–73.

Burgess, J., Limb, M. and Harrison, C.M. (1988) 'Exploring environmental values through the medium of small groups: 1. Theory and practice', *Environment and Planning A* 20: 309–26.

Catling, S.J. (1979) 'Maps and cognitive maps: the young child's perception', *Geography* 64: 288–96.

Dijkink, G. and Elbers, E. (1981) 'The development of geographic representation in children: cognitive and affective aspects of model-building behaviour', *Tijdschrift voor Economie en Sociale Geografie* 72, 1: 2–16.

Downs, R., Liben, L.S., and Daggs, D., (1988) 'On education and geographers: the role of cognitive developmental theory in geographic education', *Annals of the Association of American Geographers* 78: 680–700.

Evernden, N. (1992) *The Social Creation of Nature*, Baltimore: Johns Hopkins University Press.

Fitzsimmons, M. (1989) 'The matter of nature', *Antipode* 21, 1: 106–20.

Hart, R. (1984) 'The geography of children and children's geographies', in T. Saarinen, D. Seamon and J.L. Sell (eds) *Environmental Perception and Behavior: An Inventory and Prospect*, Department of Geography, University of Chicago, Research Paper No. 209: 99–129.

Hill. R.D. (1973) 'The vegetation map of Singapore: a first approximation', *Journal of Tropical Geography* 45: 25–33.

Jackson, L.E. and Johnston, R.J. (1974) 'Underlying regularities to mental maps: an investigation of relationships among age, experience, and spatial preferences', *Geographical Analysis* 6: 69–84.

Kong, L. and Ng, W.K. (1996) 'Beyond the classroom: the development of "place" and identity among schoolchildren', paper presented at A Place For Learning: PWD School Seminar, Singapore, 16 November.

Kong, L. and Yeoh, B.S.A. (1996) 'Social constructions of nature in urban Singapore', *Southeast Asian Studies* 34: 402–23.

Kong, L., Yuen, B., Briffett, C. and Sodhi, N. (1997) 'Nature and nurture, purity and danger: urban women's experiences of the natural world', *Landscape Research* 22, 3: 245–66.

Kong, L., Yuen, B., Sodhi, N. and Briffett, C. (1999) 'The construction and experience of nature: perspectives of urban youths', *Tijdschrift voor Economische en Sociale Geografie* 90, 1: 3–16.

Matthews, H. (1984) 'Environmental cognition of young children: images of school and home area', *Transactions, Institute of British Geographers* 9: 89–101.

Matthews, H. (1986) 'Children as map makers', *Geographical Magazine* August: 47–9.

Matthews, H. (1992) *Making Sense of Place*, London: Croom Helm.

Matthews, H., Limb, M. and Taylor, M. (1999) 'Defining an agenda for the geography of children', *Progress in Human Geography* 23,1: 59–88.

Ministry of Information and the Arts (1996) *Singapore Facts and Pictures*, Singapore: Ministry of Information and the Arts.

Perry, M., Kong, L. and Yeoh, B. (1997) *Singapore: A Developmental City-State*, Chichester: John Wiley.

Rutheiser, C. (1993) 'Mapping contested terrains: schoolrooms and streetcorners in urban Belize', in R. Rotenberg and G. McDonogh (eds) *The Cultural Meaning of Urban Space*, Westport, Connecticut and London: Bergin & Garvey.

Sibley, D. (1995) 'Families and domestic routines: constructing the boundaries of child-hood', in S. Pile and N. Thrift (eds) *Mapping the Subject: Geographies of Cultural Transformation*, London: Routledge.

Spencer, C., Blades, M. and Morsley, K. (1989) *The Child in the Physical Environment: The Development of Spatial Knowledge and Cognition*, Chichester: John Wiley.

Valentine, G. (1996) 'Children should be seen and not heard: the production and trans-gression of adults' public space', *Urban Geography* 17, 3: 205–20.

Wong, P.P. (1989) 'The transformation of the physical environment', in K. Singh Sandhu and P. Wheatley (eds) *The Management of Success: The Moulding of Modern Singapore*, Singapore: Institute of Southeast Asian Studies.

INDEX